Environmental Management Systems
in Local Public Authorities

Contemporary Problems of Modern Societies

Edited by Peter Nitschke
and Corinna Onnen-Isemann

Volume 5

PETER LANG
Frankfurt am Main · Berlin · Bern · Bruxelles · New York · Oxford · Wien

Martin Jungwirth

Environmental Management Systems in Local Public Authorities

A Comparative Study of the Introduction and Implementation of EMAS in the United Kingdom and Germany

PETER LANG
Internationaler Verlag der Wissenschaften

Bibliographic Information published by the Deutsche Nationalbibliothek
The Deutsche Nationalbibliothek lists this publication in the Deutsche Nationalbibliografie; detailed bibliographic data is available in the internet at http://dnb.d-nb.de.

Zugl.: Vechta, Univ., Diss., 2010

Va 1
ISSN 1867-609X
ISBN 978-3-631-59531-2

www.peterlang.de

Preface

This work has been a result of a long research process that began after my internship at the European Parliament in autumn 2001. While I had the chance to get an insight view of the processes on the European level, I began to ask myself to what detail the policy ideas would come to an effect at the local level. With the peak of the EMAS discussion in Germany at that time, the main policy instrument for my research project was found.

During the years, this project took its time. I would like to thank the following people for supporting me all the years: Prof. Dr. Peter Nitschke, my scientific supervisor, for giving vital inspiration and numerous helpful inputs all the time, Dr. Ines Freier, for their fruitful discussions about EMAS and its effects on public administration, Prof. Dr. Martin Müller and other friends from the DNW network for their advice at the very beginning, Martin Schwarz, for his mental support even after long hours of work, Prof. Dr. Helga Kanning and Prof. Dr. Hubert Heinelt for their critical review at the end.

Additionally, I would like to thank the interviewees both in the UK and in Germany for their openness to talk to me about the EMAS proceses within their organisations. This is equally true for numerous experts with whom I discussed my views and ideas.

Jennifer Thomalla and Christine Neumann were supportive in preparing this text, while Thomas Böge was helpful with the two maps within this book.

For the most, I have to thank my whole family for their long and continuous support during all the years. Without their help, this work would not have been possible.

I hope that the interested reader can benefit from this book. Comments are more than welcome.

Buchholz, January 2011

Table of Contents

List of figures

List of tables

It is not only what we do,
but also what we do not do,
for which we are accountable.

Molière
(1622-1673)

1 Local Authorities between Modernisation and Europeanization: How public authorities meet the challenge of introducing a management-orientated policy instrument

1.1 Introduction

The task of this chapter is to provide a general introduction to the topic of the study followed by an outline of the research question and, finally, a presentation of the course of work. The first part of this chapter concentrates on the importance of the study for political science in general and policy research especially. It will position this research paper within the current policy research debate, focusing on programme evaluation as a method for the future design of policy instruments. The second part will elaborate on the policy field of European environmental policy with a brief discussion of the development of this policy field and the relevant policy instruments which will provide the frame for the function and importance of EMAS, the policy instrument which is at the centre of the analysis. The third part will give an overview on the polity arena in which the policy EMAS instrument is used, focusing on the essential importance of modernisation of the regional public sector and thus giving an idea of the broader use of EMAS as an instrument of public authority reform. The fourth part concentrates on the research questions which are leading this work while the fifth will give an outlook of the course of the study.

1.2 Policy evaluation as an instrument for effective governance

Since the 1950ies, the question of the effectiveness of political decisions has been posed, when the interest in political science changed from questions of power and influence towards the effects of political decisions resulting in the rise of the outcome-orientated policy analysis. This was reflected by the idea of linking practically-orientated policy analysis and academic political science to get evidence-based data for efficient policy decisions. Lasswell (1951) was one of the first to demand a policy field analysis that should be content orientated, trans-disciplinary, orientated towards problem-solving and of explicit normative nature. The underlying idea was the hypothesis that the divisive issues between political actors and community interests could normally be decided on a better level through science-based systematic analysis (Wollmann 2003, p. 338-339). One result of this was the development of the programme evaluation (Rist 1990) which can be divided into at least three subcategories. While 'policy research' is an overall term for the analysis of the policy cycle (Jann 1991), 'implementations research' is a method used to analyse the processes of putting a policy into practice (Wollmann 1991, Mayntz 1993), 'evaluation research' assesses the results of a policy as well as the main factors for success (Stockmann 2000).

While critics of policy analysis merely denote it as a socio-political trend of the time (Greven 2008) as well as describing the practical effect of policy research projects as minimal (Lindblom 1990) and even note that

> "[...] the idea that a political system is no longer governed by strategy but by communication based on agreement is the contemporary variant of the platonic utopia" (Greven 2008, p. 32, translation by author),

there is definitively a need for policy research. Schneider (2008) argues that science-based decision making is superior against all other forms of knowledge and therefore will lead to policy decisions that have a better basis:

> "Despite the fact that social science knowledge bases do not have the precision and resolution in detail for the analysis of control and regulations techniques, e.g. compared to technical or biological cybernetics, social science has acquired a large arsenal of methods like the combination of statistical analysis, detailed case studies and field experiments, the implications of numerous policies can be tested in the view of evidence-based policy science." (Schneider 2008, p. 68, translation by author)

One field of policy analysis is policy learning, based on the assumption is that there is a learning actor that is able and willing to make changes while learning (Bandelow 2003, Toens/Landwehr 2008). These ideas can be seen in contrast to the political-economic ideas of the advocacy-coalition approach (Sabatier 1993). Since the mid-1990ies, concepts of political learning have been introduced to political sciences that consider transnational policy transfer or policy diffusion and the convergence of policies (Rose 1991, Goldsmith 2003, Holzinger/Jörgens/Knill 2007). Often, these studies had a rather broad view, describing the overall effects of policy transfer and diffusion. Policy analysis can benefit from the discussions about policy learning if it can prove the learning progress more clearly (Toens/Landwehr 2008).

This study wants to contribute to the policy learning debate – but not primarily on a transnational level. The micro-dimension of the processes within organisations, using an EU regulation on a voluntary instrument of environmental policy, the "Eco-Management and Audit Scheme" (EMAS), will first give an insight into the effects of European norms on the micro-level of the European Union. Second, the comparison of the use of the EMAS regulation on the regional level, in two different administrative and legal settings – the UK and Germany – will give detailed information on the overall effects of this regulation on the national (meso) level. Third, the analysis will provide results for the revision of the EMAS regulation in view of the special needs of public administrations in general and thus contribute to the EU (macro) level of environmental policy and governance. Broadly, it is an evaluative study of a policy tool used in a specific setting in order to enhance successful European environmental integration governance.

1.3 EU environmental policy: From silence to salience

Compared to other fields of EU policy, environmental policy is relatively new. When the EU was established in 1957, or to be more precise, its predecessors, the political focus was the economic union; its goal was to create economic wealth, political stability and, in its final consequence, long-lasting peace for all member states and their citizens. The economic and environmental problems of the member states were recognised as early (or as late, some might say) as in the 1970ies. While the European institutions started their environmental activities in 1972 with a task force for environment (Bungarten 1976) which later was transformed into a 'Direction General' and is today called 'DG (Direction General) Environment', the legal status of the environmental policy was provided only through the Common Market regulations for the harmonisation and standardisation of products and goods within the Economic Union. Consequently, the environmental policy often came into conflict with economic interests.

During the first years of EU environmental legislation, the main activities for environmental legislation were motivated economically. First, the different standards for products, e.g. the amount of lead in petrol or the emission standards of cars (Holzinger 1994) were regarded as restraints for the common trade market; policy was not orientated towards a coordinated or goal-orientated European environmental approach (Jordan 1999, p. 3). Second, the growing environmental problems that became visible during the

1960ies and 1970ies, such as transnational pollution caused by industry as in the case of the river Rhine or the fact that air pollution (Caspari 1995) demanded supranational policy, needed to be tackled. Third, the European goal of the harmonisation of living conditions as given in the preamble and the article 2 of the EEC treaty was not only seen as a quantitative but also as a qualitative goal to increase the standard of living for all people in the community (Holzinger 1995, Jachtenfuchs 1996, Demmke 1994), also resulting in an improvement of the environmental conditions for the people of the community.

The growing activities demanded an overall regulation of this new field of policy that was achieved with the Single European Act of 1987. Finally, the environmental policy was taken up into the European treaties, the first being the Maastricht treaty. Here, the environmental policy was introduced into the central EU policy documents. The policy field was set up on a new basis with relatively clearly defined tasks and goals (art. 174 of the Treaty of the European Union, TEU):

"Community policy on the environment shall contribute to pursuit of the following objectives:

- preserving, protecting and improving the quality of the environment;
- protecting human health;
- prudent and rational utilization of natural resources;
- promoting measures at international level to deal with regional or worldwide environmental problems."

Over the last decades, the European environmental policy has developed tremendously. Within thirty years, "environmental policy had moved from silence to salience" (Weale 1999, p. 40). These developments can be divided into three main phases: The initial period from 1972 to 1987 can be described as the test phase of what was to come. The legitimacy for environmental policy derived primarily from free trade and common market principles. Despite the weak executive legitimation, the common actors developed an ambitious policy with targets that could not be fulfilled (Rindermann 1992). The second phase from 1987 to 1992 can be described as a consolidation of the policy field through a more detailed legal integration of the policy into the European policy process. In addition, the EU's policy programmes expanded and developed further. The European Commission developed a "pragmatic, incrementalist approach" (Rehbinder/Steward 1985, p. 246) for its environmental policy at that time. Additionally, environmental policy started to be orientated towards the following principles: The precautionary principle, the principle of origin, the polluter pays principle and the principle of integration. Together with environmental subsidiary principles and the "better" principle, the main guidelines for this policy field were defined. The third and still present phase since 1992 has been characterised by two reverse trends. While there has been a gradual increase of the legal regulations that began with the Single European Act (SEA) and went on with the treaties of Maastricht and the following, there has been a decrease in the implementations dynamic as well as a growing deficit of execution within the member states (Weale et. al. 2000, Jordan 2002, Knill 2003).

As a result of more than thirty years of EU environmental legislation, the environmental aquis communitaire now comprises over 500 legislative items (Jordan 2002). The European Commission is active on all relevant fields of environmental policy and proclaims it to be a horizontal policy for all its activities. It is without doubt that most of the environmental problems of the present time are international problems and therefore solutions can and often have to be found at the supra-national level.

Since the beginning of its first environmental action programme[1] in 1972, the majority of EU environmental policy regulations have been designed as mandatory rules of command-and-control (Knill 2003, Perkins/Neumayer 2004) leading to a relatively strict implementation need for the member states.[2] Environmental regulation was considered as an intervention – often against the policy of one or more member states. The instruments were designed to

1 The European Commission developed its own style for its environmental policy with the introduction of environmental action plan(s) (EAP), which define the policy goals over a fixed period of time and give a strategic outlook on the issues and instruments the Commission wants to develop. Despite the fact that these plans are not legally binding and can be described primarily as a political statement of intent, they have played and still play a major role in the definition of the strategic and political targets of the EU's environmental policy field (Knill 2003). Over the last decades, it has become clear that the action plans are strongly influenced by the current environmental political Zeitgeist of each period, reflecting the overall progress of the environmental policy from after care to sustainability (Wepler 1999, Knill 2003).

2 Typical examples for this policy are the EU regulations on air pollution control or water quality. Here, the European Commission defined strict limit values that were orientated towards the principles of the best available technology (bat). In this way, the EU did not only set the policy goals but also determined how to achieve them. The room for national actors became relatively small with this policy; they turned into mere implementations managers (Knill/Lenschow 1999, Knill 2003).

tackle visible pollution problems; their main focus was remedial rather than preventive action.

Beginning in the 1980ies, highly regulated industries like the chemical industry campaigned for market-based instruments and voluntary agreements as alternatives to the command-and-control mechanisms. This campaigning can be seen in connection with the neo-liberal tendencies within the Anglo-American world (see chapter two for the effects of neo-liberalism on public authorities in the UK). The political attractiveness of the instruments that were developed in the late 1980ies and early 1990ies, the so-called "New Environmental Policy Instruments" (NEPI's), began to increase in many OECD countries (Jordan et. al. 2003). NEPI's are policy instruments that are particularly market based and/or voluntary and do not have the traditional command-and-control structure or which at least retain a relatively high flexibility within their implementation. These new instruments are distributed by policy transfer. It is a process

> "[...] by which knowledge (about NEPI's) at a particular time and place is used at another time or place in a different governance setting" (Jordan et. al. 2003, p. 556)[3]

This process can be characterised by different innovation strategies or arenas.

During the 1990ies, a growing number of implementation problems were visible to the European Commission and the scientific community. Up to that time, environmental policy was dominated by the emissions-orientated technology solutions in accordance with the best available technology. Through the annual reports by the European Commission on monitoring the application of EU law, it became clear that the top-down approach was increasingly less successful with the member states. Overall, a report stated that ninety-one per cent of the Community's directives had been implemented in 1995, with some states having more than twenty per cent of all regulations in deficit. Furthermore, there were 265 suspected breaches of Community law; in 1996 over 600 complaints and infringement cases were outstanding against member states. In addition, eighty-five cases awaited a decision by the European Court of Justice (European Commission 1996, p. 2, Knill/Lenschow 2000, p. 5). This lack of implementation also had political implications because it was felt

3 NEPI's were also introduced in other parts of the world, e.g. Japan or the US (Jordan et. al. 2003, p. 4).

that this might have an overall effect on the legitimacy and credibility of the integration process (European Commission 1996, p. 6).[4]

It was also felt that the command-and-control mechanisms of the EU environmental regulations had limited success in reaching defined objectives. Furthermore, the traditional environmental law was divided into different sub-policies regarding water, soil and air etc. – they all did not have an overall problem-solving perspective of more complex policy issues as well as a precautionary vision. In addition, this traditional law was not able to

"[...] foster proactive and innovative behaviour but rather produced defensiveness on the part of those affected" (Töller/Heinelt 2003, p. 24).

What is more, policy makers declared that

"[…] they would never achieve sustainable development […] by regulations alone" (Jordan et. al. 2003, p. 13).

Another reason for the development of NEPI's was their perceived strength which was proclaimed by the European Commission which argued that especially voluntary agreements (like EMAS) would encourage policy actors to develop a proactive attitude towards their environmental activities. Furthermore, these policy instruments were considered to be more cost effective and it was assumed that they would enable an easier achievement of policy goals (European Commission 1996).

As a consequence, the EU began to develop a number of NEPI's in order to simplify its legislation and to reduce the burdens for industry. The new instruments were introduced with the 1992 fifth EAP to promote the shared responsibility for the environment. Furthermore, these new instruments reflected the subsidiary principle more than the traditional policy instruments. The new instruments were supported both by industry, which was looking for less strict rules and thus for a reduction of costs, as well as by environmental NGO's which were, at the beginning, critical of NEPI's but began to support them heavily when they realised that they would have to promote these new instruments positively to achieve an effect on certain issues they were interested in after all.

4 The infringements within the sector of environmental policy were the highest among all policies within the European Union for decades. Thus, the development described in 1996 has remained on a constantly high level (European Commission 2007).

One of these new instruments was EMAS, the European environmental management standard (see chapter three for details).[5] The EMAS regulation was developed according to the competitive arena setting which is the leading model for this type of innovation: the EU creates a setting where member states compete for economic advantage in order to minimise their regulatory adjustments costs. When EMAS was initiated on the Commission level, the states that had already used its predecessor, the British BS 7750 standard, promoted it as a basis for the new EU standard. Other countries which had their own standards opposed this development.

Although the EMAS standard was developed primarily for industrial organisations, it was soon taken up by other organisations. In the UK, the idea of first analysing and then improving or rather decreasing the environmental impacts of public authorities was developed in the early 1990ies with a programme initiated by the Department of Environment (Department of Environment Circular 2/95), giving a legal basis for the use of EMAS within public authorities. From the time of the legal possibility, the first public authorities were registered under EMAS on the basis of these national supplementary regulations to EMAS. In Germany, the ordinance supplementing the national EMAS legislation (German: Umweltauditergänzungsverordnung) was the legal basis for the introduction and implementation of EMAS beyond industrial organisations in 1998. With the 2001 revision of the EMAS regulation, the management system was generally opened to other areas than industry (see chapter three for a detailed analysis of the development of the EMAS regulation).

With the introduction and establishment of the NEPI's within European environmental policy, the question of their effectiveness came up especially when the large scale intended effects of these instruments did not take place, for example in the case of EMAS (Knill/Lenschow 2000a). Due to the fact that only a small proportion of organisations in Europe use or used EMAS at all and followed by the decrease in the number of participants at the

5 Whereas the main EU environmental legislation is specific to a problem or policy field, like water, waste or air pollution control, EMAS together with other regulations is a kind of overlapping and overall legislation. The other forms of this horizontal legislation are the directive on public access to environmental information (first issue Council Directive 90/313 EEC, latest version 2003/04 EC), the IPPC directive (first version Council Directive 96/61/EC, latest version Directive 2003/35/EC) and the regulation on the EU eco-label (Council Regulation 889/92 EEC, latest version Council Regulation 1980/2000 EC).

beginning of the millennium, the question of the effectiveness of this new instrument has been raised (Clausen/Keil/Jungwirth 2002). Compared to the number of organisations that could potentially use EMAS, the overall effect of the policy instrument is very small. However, regarding the effects on the individual organisations using the system, the effects are largely positive (Clausen/Keil/Jungwirth 2002, IEFE 2005, Environment Agency 2006). Regarding EMAS within public authorities, research so far has often concentrated on the overall policy implementation of the instrument (Heinelt et. al. 2000). Yet, Emilsson/Hjelm (2004) indicated that it is necessary to analyse the processes of EMAS within the organisation in detail to get exact knowledge of the effects of this instrument. Thus, a major task of this work is to produce inside knowledge of the processes that are necessary for a successful implementation of EMAS within public authorities.

1.4 Modernisation of the public sector

Public authorities on all levels of administration have been facing large changes and modernisation phases over the last decades. These changes of administration in Western Europe in the post-war period can be divided into three main phases. The first phase from the expansion of the administration went along with centralisation from the end of post-war-time to the end of the 1970ies.[6] The second phase, in the 1980ies, is characterised by a move for consolidation together with a decentralisation of service delivery. Finally, the third phase began in the 1990ies and is dominated by two interdependent tendencies: one is the increase in participation of society while the other is a stronger orientation towards efficiency of the public sector (Naschold 2000, p. 34). This phase is still predominant. Thus, the necessity to modernise public administration arises from numerous problems and is accountable in all EU member states. These problems (or called 'problem areas' because of their larger scale) are as follows:

- decreasing revenues and increasing public dept
- taking on additional tasks without considerable finance

6 According to Hesse/Sharpe (1991), the expansion of public administration at that time was especially an increase in local public administration. The expansion was a development contrary to that of a modern industrial state where there would be a tendency for centralisation. Only two countries, Ireland and Switzerland did not expand nor decentralised their local administration compared to their European neighbours.

- increasing demands on administration: (public) civil servants who were orientated towards rules and hierarchy are changing their roles towards a customer-orientated service provider function, people are no longer applying subjects but rather service and quality-orientated customers (Spitzer 1998, p. 131)
- opposition against higher taxes and fees
- crisis of social security systems and an increase in the need for social security
- development and implementation of modern ICT systems and structures
- reinforced competitive pressure of local authorities among each other, they are facing global competition for the settlement of new industries and new jobs
- Europeanisation of politics: processes of integration, changes of power, changed ways of financing and communication in a multi-level governance system require new strategies of public administration.

Since the beginning of the 1990ies, many countries (not only within the EU, but worldwide) have been trying to work on these problems through a wide set of modernisation activities. There are three major trends of modernisation (Pollitt/Bouckaert 2000, Naschold 2000, p. 42). First, the internal modernisations of public authorities with a focus on governance by objectives; budgeting as well as a flexibility of labour organisation are predominant instruments. Second, a tendency towards democratisation of authorities, especially the change of decision-making processes towards more citizen participation as well as a shift of public tasks towards the society, which are now much more often taken over by associations, organisations or third party authorities. Third, there are efforts to introduce market orientated mechanisms of organisation like the privatisation of parts of an administration, the introduction of market testing of services or the move towards client-customer relationships and projects.

If one compares the modernisation strategies of several countries that are documented in modernisation literature, one can distinguish clusters of modernisation styles (Naschold 2000). While in the UK and New Zealand, the process of modernisation was initiated by central government which dominated it and focused on the reduction of costs and privatisation, the situation in Sweden is totally different. Here, the reform agenda was supported by both central and local government. Based on a pragmatic and consensus-orientated approach that is generally dominating the political culture of the country, the focus was to achieve a concentration of public institutions through a wide-ranging internal modernisation. In Germany however, the modernisation activities were strongly demanded by the Common Authority for Public Management Reform (German: Kommunale Gemeinschaftsstelle für Verwaltungreform, KGSt) that was also initiating project ideas. Thus, these projects were mainly driven by single institutions, i.e. local public authorities or single federal organisations. Although there was a general agreement on the need for public authority modernisation, there was neither an overall, commonly shared strategy for modernisation nor one central actor promoting these reforms at large. Consequently, the focus of the activities in Germany is on a limited internal modernisation in order to retain the general structures of administration. These activities can often be described as pragmatic but sub-critical. The different forms of modernisation and reform are due to the various national cultures of society, politics and administration.

	Process initiator	**Mode of communication**	**Activities**
UK/New Zealand	Central government	Top down, centrally organised	Reduction of costs, privatisation
Sweden	Central and local government	Consensus orientated	Concentration of public authorities, internal modernisation
Germany	Single authorities	Pragmatic, sub-critical	Limited internal modernisation

Table 1.1: Central issues of public modernisation of the UK, New Zealand, Germany and Sweden. Source: own, after Naschold 2000.

To conclude, modernisation processes within each country are largely determined by the process initiator, the mode of communication and governance as well as by the individual activities that are carried out. These nationnal specifics strongly influence the reform of public administration on national, regional and local level. Despite the national specifics, the use of management-orientated instruments and a larger focus on the policy outcomes is a general trend. Therefore, this study tries to analyse the use of an environmental management system, as the only management system of the organisation or integrated into an overall system, as one instrument of modern, management-orientated local authority governance.

1.5 Research design

In view of the described developments of EU integration, the importance of the region within European multi-level governance and keeping the different modernisation trends in local public authorities in mind, the study is designed as a comparative evaluation of the introduction and implementation of EMAS in local public authorities in Germany and the UK.

Over the last 10 years, there has been extensive research activity about EMAS from various viewpoints. There is wide research on the development of EMAS and its effects in industrial organisations (the latest are Ortmann 2002, Baumast 2003, IEFE 2005, Freier 2005). Further, the implementation of EMAS within many different kinds of industrial branches has been examined (e.g. Hellenthal 2001, Geibel 2004). There is also work on other topics of EMAS, like its role for the mission statement and the learning of organisations (Wendisch 2004). Furthermore, the comparison of environmental management systems has been evaluated (Fronek 2003). Thus, the question of the effectiveness of EMAS as a new policy instrument has been asked when discussing the idea of new policy instruments. Bouma (2000) comes to the conclusion that for industrial organisations, both new and old instruments can complement each other if legal pressure by the top-down regulations meets with an increased self-interest that is given way through EMAS. Nevertheless, the research on EMAS and its effects in local public authorities is not very widespread. One reason is the relatively low number of participants from the public sector compared to the industrial one although there are considerable numbers of participants, especially from regional public administrations, that take part; Germany is leading in figures.

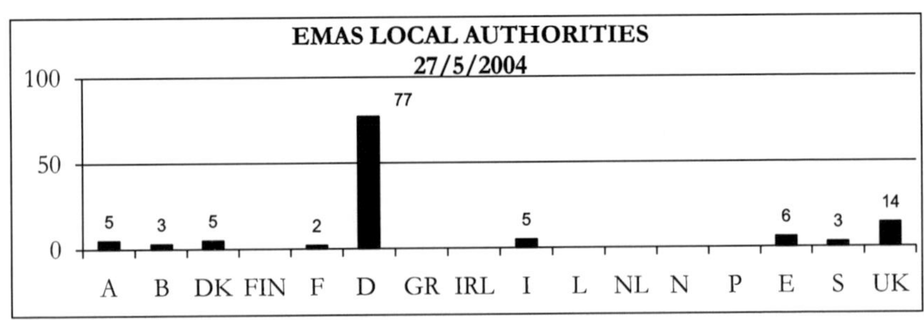

Figure 1.1: EMAS Local Authorities (NACE 75) per country. Source: EMAS Helpdesk, July 2004 (no current data available).

Generally, the overall trend with the participation of public authorities (which are registered under NACE 75) is positive with increasing numbers over the years. As with industrial organisations, there are only a very small number of authorities that use EMAS compared to the potential number of users.

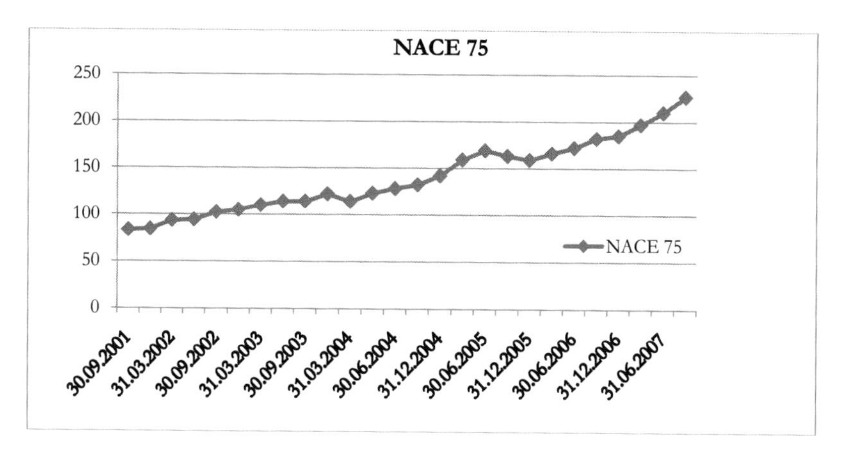

Figure 1.2: Number of organisations registered under NACE 75 2001 – 2007. Source: own, data from EMAS Helpdesk, January 2008.7

There are some case studies on EMAS in public authorities (Bentlage 1999, Funk 2000, Heinelt et. al. 2000, in parts Clausen/Keil/Jungwirth 2002, for EMAS in schools Teichert 2000) which mainly concentrate on the implementation of EMAS on the national level and focus on the administrative setting, the different actors as well as the effect and range of the implementation of the EU regulation on a national level. Up to now, there is only one detailed study which closely examines the implementation process of EMAS in public authorities (Emilsson/Hjelm 2004). The findings of the study are that there is a cultural-cognitive approach at one, a normative-regulative approach at the other authority that determines the use of EMAS within each organisation. The authors suggest that when analysing these kinds of processes, it would be wise to study the actual implementation process because this is where the actual change within the organisation begins.

Regarding practical guidelines for the implementation of EMAS within public authorities, there are a few reports and guidelines issued in Germany that

7 Since January 1, 2008, the new NACE-Code for public administration is 84.11, due to an overall NACE classification revision.

concentrate on giving advice on how to implement the system within a single organisation (Landesanstalt für Umweltschutz Baden-Württemberg 1998, Landesanstalt für Umweltschutz Baden-Württemberg 2002). There were some very early studies on EMAS in the UK which were discontinued in the late 1990ies (Riglar 1995, Riglar 1997). There are no guidelines for the work with EMAS in public administration in the UK, neither from the Department for Environment (now Department for Environment, Fisheries and Rural Affairs, defra) or the Environment Agency.

The main research questions of this work arise from the combination of EU multi-level governance, the principles of the EU horizontal legislation as well as the voluntary legislative elements of EMAS. The study wants to evaluate the results of this legislative framework within a specific form of organisation to get detailed knowledge of the outcomes of the policy instrument EMAS. Together with trends and instruments of public modernisation, it will give profound details of the effects a policy specific management system has on organisations, how this system changes the internal work (or only the wording) of the organisation and how EMAS is filled with life within each single authority using it. As local authorities proclaim and actually have a leading role in local governance, the additional question is how they can set an example in using a voluntary environmental management system.

Therefore, the main three research questions are as follows:

1. Is there a structural advantage in the style and operation of the public authorities that support the use of management-orientated instruments, in this case the introduction and implementation of EMAS?
2. How is EMAS introduced and implemented in regional public authorities in Germany and the UK? What are the main reasons for implementing it? Which policy processes lead to a successful implementtation result?
3. Which learning effects for the organisation have been achieved with the management system?

The research focuses on the processes and long-term changes initiated by the environmental management system within organisations. The environmental outcomes, as documented in the environmental statements of each organisation, are not in the focus of interest. The aim of the work is to determine whether EMAS is a reasonable, useful instrument for managing the environmental policy of a local public authority. On a broader scale, the study wants to gain knowledge on the usefulness of this kind of voluntary policy instrument of EU environmental law under the special conditions of local public authorities. Thus, it wants to contribute to the discussion of the effectiveness of context-orientated EU legislation against the immanent top-down approach that is still predominant. Nevertheless, each EU policy instrument creates its own problems and relies on the national implementation process. This is also true for EMAS as it has been perfectly implemented in the UK because it fits into the national institutional and political framework with its single case approach, while in Germany, it required national adaptation and did not correspond with the existing practice of the traditional corporatist arrangements between state and industry (Knill/Lenschow 2000a).

Furthermore, the study wants to enrich the discussion about the effectiveness of public management reform at the regional level in the UK and Germany. Taking in mind the different cultures of public authorities and especially considering NPM instruments and their long-term effects, the main question is how this instrument has been implemented and what kind of organisational effects it has had. Thus, from this point of view, EMAS is regarded as one instrument of the NPM repertoire.

With this design, the study aims to look at the processes within the authorities from two perspectives. One is from the European macro level, asking what effects EU legislation can have on the local public authorities as a special form of organisation. The second is the micro or rather intra-organisational level, asking how the implementation processes took place and what kind of effects these processes have had on the organisation.

The study is designed as an empirical implementations research project. The empirical data of the study is derived through qualitative expert interviews (see chapter five for the methodological work) that have been carried out at selected local public authorities in the UK and Germany. The selection of the authorities follows the most similar case model.

Although the policy cycle model is one theoretical basis for the analysis, it does not concentrate on mere descriptions that are often done within policy field analyses. Its major focus is the development and processing of programme innovations and the learning capacities as well as the learning blockades of the policy actors (Janning/Toens 2008 p. 12). Therefore, an analysis of the organisations' main operational structures and underlying models of public authority is the introductory part of the analysis. This is followed by the policy processes of introduction and implementation of EMAS with the policy cycle model. Finally, to get detailed information about the learning capacities and blockades, the processes are analysed with a model of organisational learning.

1.6 Course of the study

Chapter two will begin with the main aspects of public management in Germany and the UK. A historical review of the development of public authorities with a focus on local governance is given. The aim of this chapter is to explain country specifics through a historical perspective – each country's development is a reaction to the national political, economic and social setting.

After this description of the underlying operation's principles and its developments within local public authorities, the evolution and the central features of the European environmental management regulation EMAS is discussed in chapter three. Besides the political setting for the regulation, the influence of Germany and the UK on several aspects of the draft as well as the final version of the regulation is discussed. These issues provide detailed knowledge about the structural principles of politics of both countries. Furthermore, the main phases of each country's environmental policy are explained. Finally, the participation in the policy instrument EMAS and a comparison with the industrial standard ISO 14001 is laid down.

The theoretical basis for the analysis, which consists of three parts, is discussed in chapter four. As there is no consistent theory for the analysis of intra-organisational processes of the kind of EMAS, a theoretical model has been developed for this purpose which consists of three major elements. First, the two models of public authority operation, the classical Weberian Model and the New Public Management are compared. Second, the policy cycle model is described in detail. Third, the organisational learning theory is laid down. Then, all three elements of theory are combined to a new theoretical model for the analysis.

The following chapter five outlines the methods used. After a discussion about the usefulness of a qualitative approach, the method of data gathering through semi-structured expert interviews is given in detail. It is then followed by a description of the data analysis. The chapter closes with a description of the selection process of the analysed organisations as well as an overview of these organisations' main features which are relevant for the analysis.

The results of the analysis are given in chapter six. For each of the selected organisations, a detailed examination of the introduction and implementation of EMAS is given. Here, the three parts of the theory are reflected upon: first, the underlying mode of operation of the public authority; second, detailed results of the work with EMAS and third, learning processes within the organisation that are caused by the use of and work with the environmental management system. As a result, organisation-specific opportunities and problems of working with the EMAS regulation are made visible.

The conclusions of the research as well as a research outlook are presented in chapter seven. The country specific issues are followed by more general conclusions for the use of EMAS in public authorities. Then, a discussion about the effectiveness of new environmental policy instruments opens the perspective for future policy instrument design. The final part of the chapter concentrates on an outlook for further research activities around the topics of this work.

2 Public Administration in the United Kingdom and Germany

2.1 Introduction

The purpose of this chapter is to give an overview of public administration's developments in the United Kingdom and Germany beginning in the post-war era – with a special focus on local public administration. After a definition of the central term of this chapter, a brief look at the development towards the post-feudal administrative structures is given. It is followed by a country specific overview of public administrative science. After that, a description of developments of public administration organisations within the last decades in the two countries gives an overview of the major issues within the field of public administration reform. In addition, there is a description of the national developments regarding the latest reform developments, especially the introduction and implementation of New Public Management in both countries, again with a focus on the local level. Finally, a discussion of the chapter's relevance connects it with the overall goal of the study.

2.2 Central terms: Public administration, modernisation and reform

'Public administration' is a complex term because the definition inherits two major problems. One is the fact that it is difficult to include all characteristics of an administration since the concrete administration itself can be quite diverging (Ellwein 1993, p. 6). Another is that the term 'administration' is a daily phenomenon and, at the same time, a living reality. Therefore, it is easy to define it in a common and colloquial way, but it is more difficult to embrace all the aspects of 'administration' in an accurate definition (Forsthoff, 1973, p. 1). In addition, the complexity of the term is due to fact that there are numerous tasks to take in mind when analysing public administration. For example:

- the tasks and achievements of this type of organisation
- the de jure rules of work and the de facto rules of operation within a public administration
- the formal and informal structures of an administration
- the internal and external relations of a public administration

- and, finally, the staff of public organisations and questions of human resources as well as problems (Bogumil 2002, p. 4).

Therefore, a definition of the term 'public administration' is rather broadly, not as an organisation, but rather as a discipline

"[...] that is most concerned with the relationship between the state and society and with the policy process. It has its roots in the 'traditional' discipline of political science." (Osbourne/ McLaughlin, 2008, p. 70)

Consequently, the best way of describing public administration is within a concrete setting.[8] This study focuses on local public authorities, i.e. public administrations that work on a local or regional level. Therefore, the term administration here mainly considers local public authorities, unless otherwise stated.

Regarding central definitions of this work, one has to distinguish further between 'public administration reform' and 'modernisation of public administration'. Whereas 'reform' is a consciously planned, targeted changed project which will be terminated after the targets have been achieved, ‚modernisation' is an ongoing process without a definite end to it (Eichhorn, 2003, p. 1143/Nohlen 1998, p. 397 and p. 543). Both strategies of change can be used together in a way whereby single target projects that can be described as reforms are realised within a larger, long-term strategy of public administration change with is a modernisation strategy. Nevertheless, the two terms are de facto often used as synonyms.

2.3 The development towards a modern post-feudal administration

If one takes a look back into history, the administration in the sense of its current form is the result of developments that started in the nineteenth century. In the mediaeval feudal state and the class state, the sovereign rights were not differentiated functionally but depended on the changing balances of power by the leading actors. Later, even the exercise of administrative tasks by a noble peasant or an untitled bailiff can be seen as predecessor of public administration of today's type. Not until the ideas of a modern state were

8 It is also possible to describe 'administration' functionally (Weber 1980).

developed, where legislation and administration is operated in the whole territory for the whole nation, was a type of organisation as described in 2.1 developed with its main features as we still know it.

Regarding local politics, the formal introduction of local authorities began with national laws that transferred certain powers to the local sphere. In the UK, the Municipal Cooperation's Act of 1835 marks the beginning of these developments; the act allowed the establishment of directly elected municipal assemblies and gave limited power to these assemblies to decide on a number of local issues (Wilson/Game 2002). From that time on, the local public authorities were developed in the UK despite the "ultra vires-doctrine" of Parliament that was and still is in place, whereby (the national) Parliament can abandon local public authorities because there is no legislative body beyond Parliament; all decisions within the national legislation can be changed by the legislative.

In Germany, local administration reforms begun by von Stein and von Hardenberg marked the beginning of the new local autonomy, especially with the Prussian Town Law (German: Preussische Städteordnung) of 1808 (Saldern 1998, p. 23/Meyer 1998, p. 51). In contrast to the UK, this law provided a safe ground for local public administration; they were regarded as separate executive and legislative entities that were secured by law and could not be abolished by the national legislative. This situation has remained stable until the present time – local public authorities' legislative and executive rights are preserved within the German constitution (German: Grundgesetz).

2.4 Division of authority: Comparing the UK and Germany

Responsibility of an administration depends on its function within the setting of administrative organisations and their tasks and on the territorial structure it is responsible for. Within this study, the regional structures of Germany and England are important.[9] At present, Germany is generally divided into three

9 Within the EU, the member states are divided into at least three subdivisions for statistical reasons. These levels, called NUTS (French acronym for 'Nomenclature des unités territoriales statistiques', the European standard for the classification of regions) are orientated towards the total inhabitants as well as on negotiations about the size of these levels between the European Commission and the member states. Germany is divided into up to three NUTS levels, NUTS 1 being the sub-national Länder level, NUTS 2 the regional Länder level and NUTS 3 the local level. The large states are

administrative levels: federal administration, state (German: Land) administration and local administration consisting of district councils, city councils and town councils. Though the tasks of the German Landkreise and the English district councils vary, the principle is the same: providing services that the unit underneath – the German Städte and Gemeinden or the English towns and parishes are too small to provide. The two groups of organisations that will be analysed in detail within this study (see chapter five for method and sample), 'Landkreise' or shorter 'Kreise' in Germany and 'district councils' in the UK, have different tasks – due to the different structures of the local government. Nevertheless, they both have the overall task of providing services that the town or parish councils are not big enough for.

2.4.1 Local Government in the UK

In the UK, there is the central government, the administration for the four home nations (or constituent countries) as well as regional and local administration. Historically, all four nations were divided into counties. These are no longer the sole units for local administration – all four parts of the UK differ in their local administrative structure. For England, which is the relevant home nation within this study, there are currently nine Government Office Regions. These Regions are subdivided into counties and unitary authorities. Whereas the counties represent the traditional division of England and are themselves subdivided into districts, the unitary authorities combine both administrative functions of county and district. Within large urban areas, this form of single-tier authority is called Metropolitan Council. The capital is divided into Boroughs which are relatively similar to the unitary authorities.[10]
In the absence of a written constitution, local government in the UK is not secured by any law. Thus, central government has the power to change local government at any time – and has done so through several Local Government

---

divided into three levels, the small ones (Berlin, Bremen, Brandenburg, Hamburg, Saarland) only have NUTS 1 and 3. The UK is divided into up to five NUTS levels, only the first three are of relevance here. NUTS 1 level is a large regional division which is not represented by any administrative division. NUTS 2 is a subdivision of NUTS 1 and again a formal comprehension of several counties for statistical reasons. NUTS 3 is the county level.

10 Since the 1990ies, the division of counties and districts has changed in some areas. Up to that time, both levels of administration had their tasks. Because this structure was not seen as effective enough, both authority levels have merged into unitary authorities (Becker 2002).

Reform Acts; the latest passed Parliament in 2000 (Peele 2004, see also chapter 4 for details). Traditionally, local government is restricted in regard to its activity because it can work only if it has statutory authority, through a statutory requirement or through a discretionary power (Peele 2004, p. 362). Any activity that is not covered by these laws is ultra vires, i.e. beyond its powers and can be sanctioned by central government, i.e. Parliament. Despite the new Local Government Reform Act of 2000 which gave local councils more freedom to promote the local development in the areas of economic, social and environmental issues of their area, it is still largely dependent on the will of central government, especially in terms of finance.

Local government in the UK has developed differently in the four parts of the kingdom. In England, there is a system of two tiers, district and county councils, with the exception of areas that have decided to vote for a unitary authority that comprises both units and those areas that are designated metropolitan areas and, thus, have only a unitary structure. As always, London is a special case and has a third form of unitary authorities. Regarding the other parts of the kingdom, one finds unitary authorities in Wales and Scotland (called Shires) as well as in Northern Ireland.

The major distinctions between the two tiers of local government in England are their tasks, whereas unitary authorities are responsible for almost all services. The districts are responsible for (social) housing, planning application, leisure and recreation, waste collection, environmental health and revenue collection. County councils are responsible for education, strategic planning, passenger transport, highways, fire services, social services, libraries and waste disposal (Local Government Association 2006).

2.4.2 Local Government in Germany

The German Landkreise are the second tier above the town councils (German: Gemeinden). The Landkreise are secured by law (article 28 of the German constitution, the Grundgesetz). They are self-governing authorities like the town councils, meaning that they have a (restricted) autonomy from the German Länder. Landkreise are joined authorities consisting of a number of town councils, especially in rural areas. Their members are the town councils for whom they deliver services the councils are not big enough to provide. The average number of inhabitants in a German Landkreis is 177.000 (Eichhorn et. al. 2003, p. 633, von der Heide 1999, p. 123-132). In general, one can say that the town council is responsible for the local tasks while the Landkreis is responsible for tasks beyond the local reach because the provision of these services would be too costly for the relatively small town councils. Although their tasks differ among the Länder, generally these are the following: planning permission, social security, youth work, traffic security, vehicle registration and licensing, disaster prevention, foreigner's registration and administration, waste management, environmental health and nature conservation (von der Heide 1999, p. 128). Overall, the task of the Kreise is to equalise the public services as well as the living conditions within its boundaries compared to those of bigger cities.

2.5 Public Administration in the UK – An overview

Public administration in the UK is dominated by central government and thus reflects the political ideas of its current leaders. Within universities, "public administration was relatively clearly defined both as an activity and a subject of academic study" (Elcock 1991, p. 2). For many years, it was mainly a part of the study programme of politics or government degrees. From the late 1960ies onwards, however, degrees in public administration appeared, largely located in the social science departments (Greenwood/Eggins 1995, p. 143). Starting in the 1980ies, however, there was a shift from the social sciences towards management, thus reflecting the changing structures, ways of operation and the influence of management instruments in public administration (Greeenwood/Pyper/Wilson 2002).

Based on traditions that started in 1854 with the introduction of the merit principle in public administration rather than the patronage, public administration developed as a classical example of the Weberian model of administration (Richards/Smith 2003, p. 45-46) with all elements of bureaucracy. Regarding public administration organisations, with the end of the Second World War, they were concerned with the so-called "tomb to womb" policy (Stevens 2003), i.e. with the provision of a large number of public services. The state developed into a Keynesian Welfare State (KWS) that directed economic and social policies through Keynesianism and welfare policies (Jessop 1994).

From the end of the Second World War, the "gas and water socialism" (Stevens 2003) lead to the nationalisation of service providers for gas, water and electricity. Until the 1970ies, the central state was the major provider of a large number of services (Hogwood 1992). Further, economically and politically important sectors of industry, such as steel production and car factories were nationalised. The KWS model became more and more difficult in the 1960ies and 1970ies due to overall economic problems in the UK. At the same time, the administration itself was heavily criticised and seen as being elitist and unresponsive to political direction. Additionally, the Civil service was seen as not having the right intellectual and managerial qualities for running a modern state. The large state intervention as well as increasing international competition raised questions about the quality of civil servants.

Like in other countries, the discussion about reforms took place regarding the structures and size of local government as well as regarding the services that should be delivered both at national and at local level. This reform process was carried out through the work of Royal Commissions that suggested changes which then had to be decided by Parliament. Leading to a similar development to the one in Germany, i.e. to the question of the effectiveness of state activities, a commission was set up by the then prime minister Wilson to examine the structure, recruitment, management and training of the civil service, though not the relationship between civil service and ministers and the machinery of central government. In 1968, the so-called "Fulton Report" (Hennesy 1989, p. 194-208) which was part of a general modernisation of the state as well as the economy, made recommendations on recruitment, training, career structure and the general structure of government. This report was highly controversial. Due to a decline of interest in the subject of civil reform in Cabinet, the proposals were only partially implemented (Richards/Smith 2002, Peele 2004, pp. 179). Thus, the 1960ies were also regarded as the "high point of deformed modernism" (Richards/Smith 2003, p. 60), culminating in the so-called overload thesis, assuming that since the Second World War the Western governments failed to deliver what was expected by the people:

> "It was argued that, inevitably, governments had failed to deliver on many of these expectations or demands, which in turn had resulted in a serious decline of public confidence." (Richards/Smith 2002, p. 88)

Regarding the local level, the 1972 Local Government Act made way for structural reforms by introducing unitary structures (the only structure of local government in Wales and Scotland, partially for England) as well as keeping the two-tier structures of local public administration (Stevens 2003), thus reducing the number of administrations drastically from 1855 in the 1970ies to 442 by 2001 (Greenwood/Pyper/Wilson 2002).

With the election of the conservative Thatcher government in 1979, politics changed dramatically. The neo-liberal approach of the government involved the following strands: reducing the size of public administration, generating and securing greater efficiency and value for money as well as enhancing ministerial control over departmental business (Laffin 2008).[11] The reforms were strongly influenced by ideas of the NPM and can be divided into three main phases:

First, from 1979 up to 1983 there were strong tendencies to reduce the civil service due to more efficient ways of undertaking tasks. In the early 1980ies, the focus shifted towards the improvement of finances and general management, thus introducing management systems for each department. As well as the Financial Management Initiative (FMI), a programme to have decentralized management and budget structures that were orientated towards targets, these instruments can be seen as the heirs of the Fulton report which introduced the principle of economy (Richards/Smith 2003, p. 108). With the establishment of a National Audit Office and a National Audit Commission, the NPM instruments started to gain influence, as did performance indicators that were introduced for most public services, central and local. Second, in the mid 1980ies, the government concentrated on the privatisation of nationalised companies and fields of activities such as water supply and sewerage. This also had effects on the labour force. Until 1990, about 800,000 employees were transferred from the public to the private sector. By 1996, twenty-two large companies had been privatised (Richards/Smith 2002, p. 115).[12] The third phase of reform began in 1987 when the government introduced market-type mechanisms on a large scale in healthcare, community care and education. This included the purchaser-provider-split, the improvement of performance management systems (management-by-objective) and the agencification programme that separated policy-making and service delivery, leading to a large number of administrative agencies with about seventy percent of the civil

---

11 Hennessy (1989, p. 623) notes a single phrase that was symptomatic for the Thatcherite view on public administration, given in an annexe to a Cabinet committee paper on strategy and priorities in 1979: "Deprivilege the Civil Service".

12 Pyper (2001, p. 466) defines five types of privatisation: partial conversion to a limited company (e.g. Jaguar Cars which had been a part of British Leyland, conversion in 1984), total conversion to a limited company (e. g. British Telecom 1984, British Gas 1986), disposal of government shares in a company (e.g. Cable and Wireless 1981-1985 or Amersham International 1982), breaking of state monopoly, (e.g. 1983 Energy Act permitting private generation of electricity) or the injection of private provision within the framework of public service (e.g. through CCT, Best Value or PFI's).

service's workforce, the so-called Next Steps Agencies (NSA) (Pollitt/ Bouckaert 2000, Richards/Smith 2002, Peele 2004). In the 1990ies, several of these agencies were abolished due to programmes of market testing or management buy-out (e.g. Her Majesty's Stationery Office (HMSO) or HM Treasury, Jones et. al. 2001). Since then, the focus has been on customer service and delivery as well as on performance (Best Performance, Value for Money) and privatisation.

The Labour Government of Tony Blair inherited a state with some unresolved problems. After eighteen years of Conservative governments, it became obvious that a total market economy orientation did not solve all the problems. At the same time, these markets produced a number of unacceptable social outcomes. Consequently, the Labour party adapted the policy concept of the "Third Way" (Giddens 1997), combining some reforms that had been begun under the Conservative governments as well as taking up its roots of Keynesian welfarism. On the one hand, the government addressed social problems (e.g. problems of social order, family breakdown, welfare dependence, education and health) and defined them as tasks of the whole civil society that had to be solved. On the other hand, Labour continued to use instruments introduced by Conservative governments to tackle these problems, i.e. through PFI's or PPP-projects. Thus, they changed their policy focus by establishing new incentive structures rather than establishing new administrative organisations (Richards/Smith 2002, p. 236).

Another strategy of problem solving under the Third Way is the participation of stakeholders. This model is widely used in the education sector, but also within government departments like the Department for Environment, Fisheries and Rural Affairs (DEFRA):

> "Policy affecting air quality potentially affects many stakeholders and the departments used several methods to consult them about the strategy". (National Audit Office 2002, p. 4)

The key to these stakeholder networks and their activities is trust, whilst no formal structure is given beforehand. For complex problems, joint solutions were established as indicated in the White Paper "Modernising Government" (HMSO 1999) which listed the major issues of good policy making of the Labour government, a policy that was strategic, holistic and focused on outcome orientation. At the same time, it was evidence based, focussing on clearly defined objectives (HMSO 1999, Richards/Smith 2002, p. 241). Currently, there is no evidence that the new Labour government under Gordon Brown is changing these policy principles. Nevertheless, Laffin

(2008) describes that the 'joined-up' approach is not working and that there are numerous policy inconsistencies in central government's approach to local government. Stoker (2004) argues that these inconsistencies are due to a strategy of 'government by lottery' leading to the effect that local authority actors are destabilised in order to establish new ways of working.

2.6 The UK and New Public Management

The way public administration worked until the last quarter of the 20th century (as described above) did not seem to be appropriate any more. Instead, the administration came under financial and political pressure:

"Characterised by rule, regulations, routine and a large and expanding career staff, bureaucracy was increasingly seen as unresponsive, costly and inefficient." (Greenwood/Pyper/Wilson 2002, p. 9-10)

With the development of Thatcherism[13], NPM got a boost in the UK. Its elements were implemented both at national as well as at local level. It was a top-down process initiated by Whitehall and involved all parts and levels of public administration with a large number of instruments, ideas and activities. Based on the public choice theory and the New Right movement, the introduction of NPM started in the UK. The first activities were considered to be a change of administration's operation towards a flexible market-based form of management through compulsory competitive tendering (CCT [14]), which was a mayor driver in the 1980ies, followed by contracting out and market-testing in the 1990ies. A second instrument was privatisation. The sale of public companies between 1979 and 1996 generated about sixty-four million GBP. Other methods used to generate finances for the public sector were Private Finance Initiatives (PFI's) where infrastructure was built by private investors

13 A political string named after Margaret Thatcher, UK's prime minister from 1979 to 1990, which can be characterised as a free-market economy, a reduction of state activities, the increase of privatisation and the reduction of the state towards its core functions (Jones et. al. 2001 p. 586.).

14 This also includes tendering against in-house departments which themselves will only get the tender if they have a competitive price compared to others (Wegener 2002). Not surprisingly, a lot of contracts were won by the administration itself, so it seems that they were more competitive than the Thatcher government had thought. Furthermore, CCT was in parts very unpopular among local government managers. They were very energetic to find ways around it (Löffler 2003, p. 85).

and leased back to the state. Compared to the pre-Thatcher years, over fifty per cent of the public sector had been placed in private hands by the 1990ies (Isaac-Henry et. al.. 1993, p. 6).

Due to the large number of agencies (so-called "Next Steps Agencies", NSA) and the large number of quango bodies that have responsibilities within the network of organisations, central government set up a system for regulation and inspection, notably the Audit Commission, which is responsible for auditing the NSAs and the local authorities. The large number of agencies led to a fragmentation of services[15] which was also due to contracting out which led to a client/purchaser and/or contractor/provider split. These developments were partially criticised because it was stated that the state would lose its functions and would become hollow (Rhodes 1994). Furthermore, central as well as local governments tended to become mere administrators of contracts and become increasingly dependent on their service providers. Consequently, this sometimes leads to a paralysing situation of the organisation which gives a service to a contractor if the latter has difficulties in providing it.

The introduction of consumerism, the ability to choose services (slogan: "The right to choose") and the idea of "delivering" services for people were further steps in this direction.[16] All these developments are described by some analysts as a "Post-Fordist service delivery pattern" (Stoker/Mossberger 1995) focused on the most individual needs of people rather than the "mass production" of standardised service delivery of the Keynesian Welfare State.

For public authorities and quangos that took up public tasks, performance management was a large task within the reform processes. Based on the "three E`s - efficiency, effectiveness and economy", the transformation towards management accounting in the public sector began. At first, the 'Value for Money' principle was introduced, meaning that the lowest price was the best, regardless of the quality of the service or the related issues like work conditions, etc. Underlying this principle was the question whether private sector instruments could help the public administration to become more efficient and how these instruments could be transferred to the public sector (Greenwood/Pyper/Wilson 2002, p. 8).

15 One of the worst examples is British Rail, which was split into more than 100 organisations, then privatised and later partially turned back into a public company.

16 Thus, it has to be noted that there are limits to consumerism such as in policing or prison service. There, people will never feel like customers.

The Labour government continued this policy with its idea of the "Third Way". One instrument which is used in large infrastructure projects are Public-Private-Partnerships (PPP), of which the most prominent example is the current and ongoing modernisation of the London Underground. Regarding the local authorities, the government

> "[...] has continued the Conservatives' stress on a mixed economy of provision at the local level, casting authorities in an enabling rather than delivering role" (Laffin 2008, p. 112)

The initiatives that were focused on the local level are commonly known as the 'Local Government Modernisation Agenda' (LGMA) (Downe/Martin 2006).

Under Blair and within the context of the modernisation of local government, CCT and the 'Value for Money' principle were replaced by the 'Best Value' principle (BV) (Department of the Environment, Transport and the Regions 1998). The main reason for the replacement of CCT was that it proved to be extremely unpopular among Labour-run councils (Downe/Martin 2006, p. 446). Under this scheme, each authority has to assess its priorities in order to draw up a programme for a performance review which has to be carried out according to the following four elements (the so-called '4Cs').

- challenge why and how a service is being provided
- compare one's own organisation's performance with that of other authorities, according to a number of relevant indicators
- consult with local taxpayers, service users and the business community about the setting of the new performance targets and
- use competition as a means to get efficient and effective services (Boyne 1999, p. 6, Greenwood/Pyper/Wilson 2002, p. 143).

The focus was drawn to the competition element. Thus, the government did not want everything to be put on tender (Department of the Environment, Transport and the Regions 1998). The 'Best Value' scheme was well received by local government because authorities saw it as less prescriptive than the earlier CCT (Laffin 2008, p. 114).[17]
In 2001, Whitehall introduced the Comprehensive Performance Assessment (CPA) which superseded Best Value. While BV was concentrated on single issues, CPA's focus is on the whole performance of an authority. An organi-

17 The so-called "Best Value Performance Indicators (BVPI's)", the performance is documented in a "Best Value Performance Plan (BVPP)". Details and BVPI reports of every authority can be found at www.bvpi.gov.uk.

sation's performance is measured and compared to other authorities with the help of performance indicators which are considered a rational planning instrument. Set by Whitehall, these indicators have a large impact on the work of each authority. The performance is, as with BV, policed by the Audit Commission which undertakes regular inspections of the performance of each authority or service. The publicly available outcomes of these inspections have a punitive element in them because the poorest-performing authorities face external intervention by central government to recover according to national performance standards. The incentives connected to CPA are limited; even the best performing authorities have received little of the additional freedom promised by central government (Sullivan/Gillanders 2005). Nevertheless, CPA has been successful over the last years. By 2005, the Audit Commission had judged almost three quarters of unitary and upper-tier local government as 'good' or 'excellent' although the overall performance of local government was declining, as was public satisfaction (Martin/Bovaird 2005). Interestingly, the results of the audits are improving the image of local government because they are seen in a sharp contrast to the management failings of central government's departments (Laffin 2008, p. 115). The overall impression regarding local authorities' modernisation is that central government is driving change while the local level is remaining passive in activity. This passive and defensive role towards the centre is historically the norm rather than the exception (Laffin 2008, p. 116).

To conclude, one can say that the UK has introduced many of the ideas and concepts associated with NPM. It has led to a performance-driven, competition-orientated public administration both at local and national level. It has changed the style of operating administration dramatically. British administration today is on the whole management-orientated and focused on efficiency.[18] Nevertheless, there is a conflict between competing cultures. On the one hand, there is the traditional Whitehall culture with its public service ethos and the idea of a hierarchically-organised government. On the other, there are the NPM instruments and the idea the administration should be merely working on policy implementation whilst being as close as possible to the market economy (Richards/Smith 2002). The idea of a more joined up

18 Despite the general trend, one can see some negative trends of NPM in the UK. A good example is the problem of social exclusion. Due to the new orientation in managing public services, the social exclusion has become a larger problem in the UK, or at least has become more visible. Therefore, it is very interesting to notice that the Labour government established two new units in Number 10 when coming to power in 1997; one being the Unit for modernizing government, the other is the Social Exclusion Unit.

work and policy making that was promoted by Tony Blair's government (Ling 2002) has not been achieved. Especially during the first years, Whitehall was unable or unwilling to coordinate the wide range of activities and initiatives because of a lack of horizontal coordination. This made it difficult to address horizontal issues. Downe and Martin (2006, p. 470) note:

"[...] the persistence of sectorally based funding regimes, performance management systems and inspectorates combined with fiercely independent professional networks mean that after nine years of local government 'reform', local agencies still struggle even to share, let alone coordinate their actions, sometimes with disastrous results [..]."

Despite this, the wide range of activities under LGMA all contributed to the overall outcomes of the following key objectives: service improvement, community leadership, accountability, stakeholder engagement and public confidence, despite the fact that local practitioners claim to be overwhelmed by the volume of the initiatives and the speed they have to implement them (Downe/Martin 2006).

2.7 Public administration in Germany – An Overview

In Germany, public administration science, the so-called 'Verwaltungswissenschaften'[19], was strongly dominated by the new orientation of German Political Science after the Second World War. Additionally, the German perspective was dominated by lawyers who saw public administration from their own deterministic professional background due to the German 'rule of law' principle (German: Rechtsstaatsprinzip). Thus, in the 1970ies, the perspective changed because questions of political outcome became prominent. In addition, great changes in Germany's political and social systems, questions of planning political programmes as well as of output and outcome lead to an increasing interest in public administration science.[20] From that time on, social

19 The German term 'Verwaltungswissenschaft' sometimes used in plural form as 'Verwaltungswissenschaften' to show that it is not a genuine discipline, but incorporates several strings of science in Germany, does mean Public Administration but is often used for the term Public Policy as well (Bogumil/Jann 2005, p. 34). Sometimes the term 'Verwaltungslehre' (science of administration) is used to show that this is a more practical approach, especially in regard to the education of public administration staff (e. g. Püttner 2000).

20 Another important term is „Verwaltungspolitik" (politics of administration), meaning the guidance by the legitimate leadership of the administration to develop, implement and control of the substance, procedures and styles of administration as well as the

and political scientists were needed and started to work within the field of administrative science.

Based on the findings that politics are always a method to actively solve problems of the society while its processes tend to be increasingly complex and difficult, the question of political control gained a larger interest. In turn, this affected public administration as well as public administration science (Bogumil/Jann 2005, p. 32). Thus, there was and still is a research focus on the administration as an organisation itself and the administrative side of policy making. This is reflected in the term political-administrative system (PAS) (Grauhan 1969, Offe 1972) which does not only break with the tradition of the American Public Administration that determined a role where the administration has a purely executive function. Additionally, the idea of the political-administrative complex as a system turns away from the idealistic idea of the division of powers of traditional political science. Consequently, the new term PAS opened up the horizon for the idea that administration is per se always involved in policy-making.

Empirical research revealed that a centralistic, mono-rationalistic hierarchical public administration was unrealistic and therefore suggested considering a more complex view of public administration which is characterised by numerous actors, a large number of rationalities and networking among the relevant organisations. This led to a change of paradigms as from then on the state was no longer seen as the centre of social regulation. There was a growing interest in the characteristics of political sub systems and in policies and programmes like environmental policies, labour policies and economic policies rather than on polity like government and administration itself. The theoretical basis for this was the actor-centred institutionalism (Mayntz/Scharpf 1995, Czada 1998, Scharpf 2000) that considers the dynamics of development of society and concentrates on the interaction of corporative actors, policy networks and systems of negotiation as systems of political steering. The intra-organisational perspective is less interesting and, as a result, policy research with a focus on public administration itself decreased in importance. This lack of focus on the internal effects of Public Administration was filled in the 1990ies with concepts of business administration which were summarised under the term 'Public Management' or 'New Public Management' (NPM).[21]

---

management of the organisational and staffing of the organisation (Böhret 2001, p. 43).

21 Compared to other countries, the concepts of NPM were picked up lately due to a lack of interest that had its reason in the relatively positive situation in Germany in the

From around the year 2000, the Public Administration science developed the concept of governance (Jann 2002, Richards/Smith 2002, Benz et. al. 2003, Benz et. al. 2007) which comprises all forms and mechanisms of co-ordination between more or less autonomous actors which interact interdependently (Benz et. al. 2007, p. 9). This concept has a multiplicity of meanings. One is the sensitization of the numerous actors, the terrains and the relationships involved in governing. The other is the implication that the traditional role of government has been changed to the extent that it is now one actor among many in policy processes. Consequently, according to the governance concept, there is a pluralisation of power.

"Put another way, power has been dispersed away from the traditional central-state actor to many different and new arenas and now includes many, often new, actors within the political system." (Richards/Smith 2002, p. 19)

In Germany, federal administration is responsible for service law which has its effects on all levels of administration. Furthermore, through the transfer of its tasks to sub-national units, it affects these as well. Public administration reform processes are primarily implemented by the Länder; these are responsible for the largest number of administrative tasks. The Länder independently organise their administration and set the frame for the local authorities through their local community law. The local administration also has its own right to organise itself according to the local needs. Thus it has a large autonomy with its own tasks as well as with the ones it has do carry out on behalf of the Länder or the Federal administration (Nassmacher/ Nassmacher 1999, Bogumil/Jann 2005, p. 192).

Regarding the historical development of public administration reform in Germany, primarily the period from the end of the Second World War is of interest within this work. Leading actors, among them the Allies, as well as national and local German politicians were interested in rebuilding a working administration on all levels. However, there was no stringent or overall plan nor was there a unitary policy for this rebuilding process. Administration was intended to continue as before the war and the Third Reich. Consequently, there was an almost path-dependent re-orientation towards the administrative structures and procedures of the Weimar Republic (Hesse 1990, Wollmann

---

1980ies. Additionally, it was due to some specifically German regulation like the federalist system of states, the guarantee of local self-government and the de-concentration of administrative functions that for a long time gave a lead in modernisation in Germany (Wollmann 1996, p. 19).

2000). Nevertheless, there were both national and sub-national plans for the simplification of administration (Wagener 1979). Most of these activities were of little effect. Some adaptations took place but the main elements of administration remained untouched. Thus, administration remained shaped by the hierarchical and legal-bound rule of the Weberian model of administration (Wollmann 2000, Bogumil/Jann 2005).

The time from the mid 1960ies until the mid 1970ies can be seen as the end of the post-war reconstruction period. The concept of a pro-active and interventionist state led to a Neo-Keynesian perspective. Thus, an interventionist state model was developed that would require large capacities of planning as well as an expansion of infrastructure. The state should be able to develop forward-looking politics and be able to correct and or even predict the failures of a market economy (Bogumil/Jann 2005, p. 193). Due to increasing problems of the state in fulfilling this role, numerous modernisation processes were initiated: a reform to equalise the financial situation of the Länder, a process to reorganise the Federal Government and its structures and capabilities which was only adapted in parts (Lepper 1976, König/Miller 1995) and a territorial reform of the Länder. This reduced the number of counties and municipalities drastically (Laux 1999) but left the federal principles (of Western Germany) untouched. According to Ellwein (1994, p. 73), this territorial reform is the only reform project of public administration that has been really successful. The planned reform of the national service law did not take place due to a lack of finances, the resistance of lobby groups and the Liberal Democratic Party as well as implementation problems (Bogumil/Jann 2005, p. 194).

From the mid 1970ies until the end of the 1980ies in Germany, the public administrative system was working on processes of 'task scrutiny' (German: "Aufgabenkritik"), which was to reconsider local government's tasks through a kind of cost-benefit analysis, as well as activities to improve the citizen-orientation while, at the same time, the international discussion about NPM began. 'Task scrutiny' was a reaction to the financial pressures of the time as well as a questioning of the size of administration after its expansion in the earlier period. The citizen-orientation related to the criticism that administration worked imperso-nally towards the general public and lacked service-orientation (Mayntz 1980a, Ellwein/Hesse 1985). In a way, the German developments in public administration reform diverged from the international developments. Thus, the NPM discussion was not taken up due to national reasons. First, there was the widespread impression that the German administration performed well – also in comparison to international standards.

Second, the German conception of the state itself, which – in contrast to the Anglo-Saxon world – is distinct from both society and thus also from economy, was one reason for this rather individual development (König/ Füchtner 1998, p. 9).

At the end of the 1980ies, with a new conservative federal government let by the chancellor Helmut Kohl, political priorities changed. From the welfare state of the former governments, the new one took up the international trend of neo-liberalism. Central government sold its stakes in four large industrial companies and privatised the federal railways (Deutsche Bundesbahn) as well as the Federal Postal Service (Deutsche Bundespost), thus generating twelve billion DM (or about six million Euro). In the process, about one million people were transferred from public service into the new private companies (Bogumil/Jann 2005, p. 198). Nevertheless, the German transformation process was still dominated by the social market economy, thus retaining the high standards of social security as well as a strong participation of unions and other civil society organisations in within the market economy.

With a ten year delay, the NPM movement affected Germany in the 1990ies. This was due to numerous reasons (Schröter/Wollmann 1997). First, there was the financial situation of Germany with rising public debt due to an economic decline and the financial needs of reunification. At the same time, the Maastricht criteria asked for strict budgetary austerity as demanded by Germany itself during the treaty consultations. Consequently, the neo-liberal ideas of a small state and an economically-efficient administration seemed highly attractive. Second, there was decreasing trust in the quality of German public administration as a result of a study of performance of public administration worldwide, financed by the Bertelsmann Foundation in 1992. As a central result, the study stated that large German city administrations were rated at the bottom while cities like Phoenix and Christchurch, which were working according to NPM principles, were outperforming others (Bertelsmann Stiftung 1993). A third reason for the trend towards NPM was the change of strategy of the 'Joint Local Government Agency for the Simplification of Administrative Procedures', the local authorities' modernisation body. While until the late 1980ies, this organisation recommended a rather hierarchical and rule-bound Weberian model of public administration, it changed its position in the early 1990ies by strongly criticising the public administrations' mode of operation (Banner 1991, KGSt 1994) and promoted a new model of operation, orientated towards the NPM-model. After about ten years of NPM discussion in Germany, the current discussion within public administration science has, since the end of the 1990ies, turned partly towards a public administration that takes the surrounding society into its focus. The 'activating state' considers problems of society as well as of administration and encourages the public to participate in the developments within their local region (Bogumil/Jann 2005). Led by the federal government of Social Democrats (SPD) and Greens (Grüne), the government agreed upon a programme that balances the duties of both government and administration with the self-initiative of the people and NGOs of its society (Bundesregierung 1999, Bundesministerium des Innern 2005, Bundesministerium des Innern 2006). Nevertheless, most reform initiatives were carried out by the Länder authorities or local public administrations.

2.8 Germany and New Public Management

When NPM came up in Germany in the 1990ies, a national variant was developed: the "Neues Steuerungsmodell" (New Steering Model) as developed by the KGSt. It resembles the NPM-model but because the German discussion at the beginning, did not reflect intensively on the international discussion but took the so-called 'Tilburg Model' (Kleinfeld 1996, p. 186-217), i.e. the reform activities of the Dutch City of Tilburg as the leading pattern. Its first focus was on contract management, decentralised resource responsibility and efficiency. The main cause of problems had been seen in a number of gaps of management (Steuerungslücken) by the KGSt regarding the efficiency of public administration and internal management - especially the ability to react to a changing environment, a lack of proof of the legitimacy of used finances as well as a decreased attraction of the public sector for new and existing staff (KGSt 1993). To solve the problems of the public sector, the main task was not the reduction of instruments of social security as had been the case in many neo-liberal countries like the UK in the 1980ies, but to install improved measures of internal management (German: Steuerung) mechanisms. Overall, NSM was a new way of organising public services and reflects obvious failures/weaknesses of the present system of public administration. Its concept is based on the three following elements:

- building of corporate, decentralised management structures and capabilities
- output-orientated management and the use of equivalent instruments as well as the
- activation of these structures through competition and customer-orientation (KGSt 1993, Damkowski/Precht 1995).

Thus, the NSM (and NPM)-model is an alternative to the classical bureaucratic steering that was not capable of solving the arising problems.

During the 1990ies, NSM became a trendsetter for modernising public administration on all levels of public administration but with different intensity (Wegrich et al.1997, Gerstlberger/Kneissler 2000, Osner 2001). Whereas it has had a larger impact than previous intentions of modernisation, it is clear that the NSM is a bottom-up development in the German public administration, meaning that the largest group of public administrations that have used NSM or parts of it to change their structures is the local level.[22] There are noticeably fewer institutions at the regional/state and central level that have introduced elements of NSM as part of their reform process.[23]

Regarding the instruments used for reform, many of them are intended to develop a better use of financial resources as well as the outsourcing of services. Structural changes of larger extent are not in the primary focus. Regarding the aims that were proclaimed at the beginning of this reform period, one can say that, relatively seen, the reform has been a failure (Jann et. al. 2004, Bogumil/Jann 2005 p. 205-208). Despite its high aims, there are a number of unintended or even contrasting effects, especially in regard to the implementation of instruments taken from business administration, such as

- problems with contract management
- less bureaucracy can lead to less control
- there is no output-orientated management concept and
- the concentration on product catalogues by public administrations did not lead to the effects intended (Bogumil/Jann 2005, p. 205-206).

On the one hand, these developments and effects are caused by insufficient implementation of instruments due to the burden of financial restraints, partly

22 Reform activities after the Second World War already began in 1949 when the KGSt was created to develop strategies for simplifying procedures within local public administration. This was given up within a relatively short time; the rebuilding of West Germany and the so-called "Wirtschaftswunder" (German for economic wonder, meaning strong economic growth) was seen as more important. Despite this, the KGSt remained as an advisory organisation for local authorities.

23 Of course, the number of authorities on the local level is considerably larger than that on the state (Länder) or on the central level. Nevertheless, it is remarkable that only at the local level NSM was used to a wide extent.

conservative structures and the lack of an exit option. On the other hand, the problems are due to a misleading theory which does not take into account the rationality of political processes. With the experiences of NSM, the management style and the idea of overall efficiency for public administration, the discussion among German public administration practitioners and researchers turned to governance and civil society issues from the late 1990ies (Schwalb/ Walk 2007).

2.9 Relevance for the study

The reforms of public administration both in Germany and the UK are the starting point of the study. As described in this chapter, both countries have different structures of administration as well as different reform agendas. Both countries traditionally had a Weberian-orientated public administration. Thus, one was centrally dominated whereas the other was dominated by federalism (at least West Germany from 1949 onwards). Although the developments were different in detail, both countries have seen the phases of increasing the Keynesian welfare state followed by a phase of strong neo-liberal instrumentalism and market-orientation and, finally, a return to a stronger cooperation between administration and civil society.

In 1979, the UK began to implement various instruments of NPM in a top-down process. This nationwide reform process has influenced all levels of public administration and has fundamentally had enormous impacts on the local level. In the UK, NPM is widely implemented, thus leading to a neo-liberal, post-fordist attitude of public administration.

In Germany, NPM in its specific variant of NSM is more or less a bottom-up process which, compared to the international discussion, started late, i.e. in the 1990ies, and has lost some of its dynamics over the last years. Germany's public administration is, in general, still dominated by the Weberian Model although especially on the local level, there are reform results regarding the input side, less for the output and the process dimension. Very little has changed regarding strategy and management (Bogumil/Kuhlmann 2004).

Details of the policy instrument EMAS will be outlined in chapter three. It will become clear that it can be seen as an NPM-like instrument. Although it is not focused on a single-issue, its overall management-orientation, the flexibility of its implementation and the orientation towards performance indicators classify it as an instrument of NPM within the context of the other modern instruments and policies both German and UK local authorities are using.

3 The policy instrument EMAS: From the development on the European level towards the national implementation in the UK and Germany

3.1 Introduction

This chapter deals with the development of the EMAS regulation[24] both on a European as well as on a national level in the UK and Germany. First, a general overview of the main features of the European environmental management standard is given. It is then followed by a discussion of the political and historical development of the first regulation on the European environmental management system EMAS I. The second part of this chapter is an overview of the EMAS II regulation as of 2001 as well as a brief outline of the major changes comparing the initial and the revised version of the regulation. A third part deals with the second revision of the EMAS regulation, the so-called EMAS III, which is currently under way. Additionally, a short overview of the situation of the two most-used and competing environmental management systems, EMAS and ISO 14001, is given due to the fact that with the EMAS II regulation both EMS are interchangeable to a large extent. The fourth part of this chapter deals with the regulatory and institutional implementation of EMAS in the UK and Germany, followed by an overview of the overall development of EMAS in each country. The fifth and final part concentrates on the use of EMAS within public authorities in both countries and closes with an outlook on EMAS, both regarding its general development in regard of the forthcoming revision and the special situation of EMAS in public authorities in the UK and Germany.

3.2 Overall structure of EMAS

The 'European regulation called 'Eco-Management and Audit Scheme' has been in use since 1995 and was revised in 2001. The general structures of the scheme have not changed. Nevertheless, some detailed aspects like the site vs. the organisational view of the management system changed from EMAS I to

---

24 Although there are the regulation 1836/93/EEC (EMAS I) and the regulation 761/2001/EC (EMAS II), the term 'EMAS regulation' in its single form is used because both regulations have the same policy instrument as content and belong together.

EMAS II. The following description concentrates on the current EMAS II regulation; changes from EMAS I to EMAS II are described in part 3.5 of this chapter. In short, EMAS is a site-related environmental management system standard with internal and external validation which requires organisations to publish an environmental statement and provides a verification system with independent verifiers and an official registration (art. 1 EMAS II regulation).

The use of EMAS is voluntary for all organisations (art. 3 EMAS II regulation) that are interested in applying for the scheme. For statistical and operational reasons, these organisations are classified according to the European NACE-scheme which is controlled and supported by the European Commission. EMAS can be introduced into an organisation or into units of one and provides a framework for a management system that primarily deals with environmental issues. It does not set specific targets that have to be achieved but provides a flexible, yet demanding management framework for the organisation's activities. Politically, it is regarded as a premium environmental management standard due to its relation to EU law, the requirements of internal and external verification and the obligation to publish an environmental statement.

When an organisation begins with the EMAS process, it has to begin with an initial internal environmental review in which all environmental aspects of an organisation have to be examined closely, including direct ones like water and energy consumption, waste disposal, stationery/office supply, raw-materials, hazardous substances as well as indirect aspects like information, procurement, waste management, energy management, etc. The initial review is the basis for the management cycle that is supposed to be continuous, based on the model of the PDCA-cycle (Plan, Do, Check, Act) by Deming (1993). The first step within this continuous circle, which, according to the EMAS regulation, is due to be completed every year (although the environmental declaration has to be published every three years but updated yearly), is to document an 'environmental policy' of the organisation that has to be adopted by its senior management. This environmental policy includes a commitment to continuous improvement and prevention of pollution and a commitment to comply with relevant environmental legislation. Further, it provides the framework for setting and reviewing the environmental objectives and targets. This document is followed by a decision on a programme of activities, the 'environmental programme'. It covers a detailed plan for the single activities to be carried out in order to reduce the organisation's environmental impacts and also includes the responsibilities as well as a time frame for these activities. The environmental programme is the nucleus for the 'management programme' that covers not only the environmental programme itself but also a plan of the responsibilities and duties of the relevant staff members, a guideline of all relevant legal documents and other necessary documentation. Regularly, usually once a year, an environmental audit is performed which evaluates the management system, its compliance with the relevant EU, national and regional environmental legislation and the organisation's environmental goals. This audit is the basis for the corrective actions that have to be taken to improve the overall management system. At the same time, the organisation shall set new environmental objectives for the future which will then be documented in the revised environmental policy for the management cycle.

In order to document its environmental activities and improvements, an organisation is obliged under EMAS to publish a publicly-available environmental statement that includes a description of the organisation's activities, together with documents of environmental management such as the environmental policy, programme and management system as well as an assessment of all environmentally-relevant data and issues. Usually, the environmental outcomes and the use of resources are presented by statistics and other quantitative material, especially with the direct environmental impacts. Compared to this, the documentation of indirect environmental aspects is a complex task.[25]

The next step within EMAS is the verification of the management system, the environmental policy and programme as well as the environmental statement through an independent verifier who has to be commissioned by the organisation. Verifiers are experts in their fields and specialised in some fields of industry and/or types of organisation. They have to be accredited by a national body responsible for this process according to the EMAS regulation (art. 4 EMAS II regulation). When the verifier has checked the environmental policy as well as the management system and given a positive statement that the EMAS regulation has been met, the validation process is completed with the registration of the organisation's site that uses EMAS at the relevant national responsible body (art. 5 EMAS II regulation). The data provided at registra-

25 The requirement to document indirect environmental impacts is one major factor for organisations to quit participation in EMAS over the last years (EVER Study 2005).

tion is also given to the European Commission for its European-wide EMAS register.[26]

After successful validation and registration, the authority is allowed to use the EMAS logo and publish the environmental statement for distribution to the public. There are detailed rules for the use of the logo in public (art. 8 EMAS II regulation[27]). Thus, the logo can be used to promote the environmental activities and capabilities of the organisation but not in connection with its products.

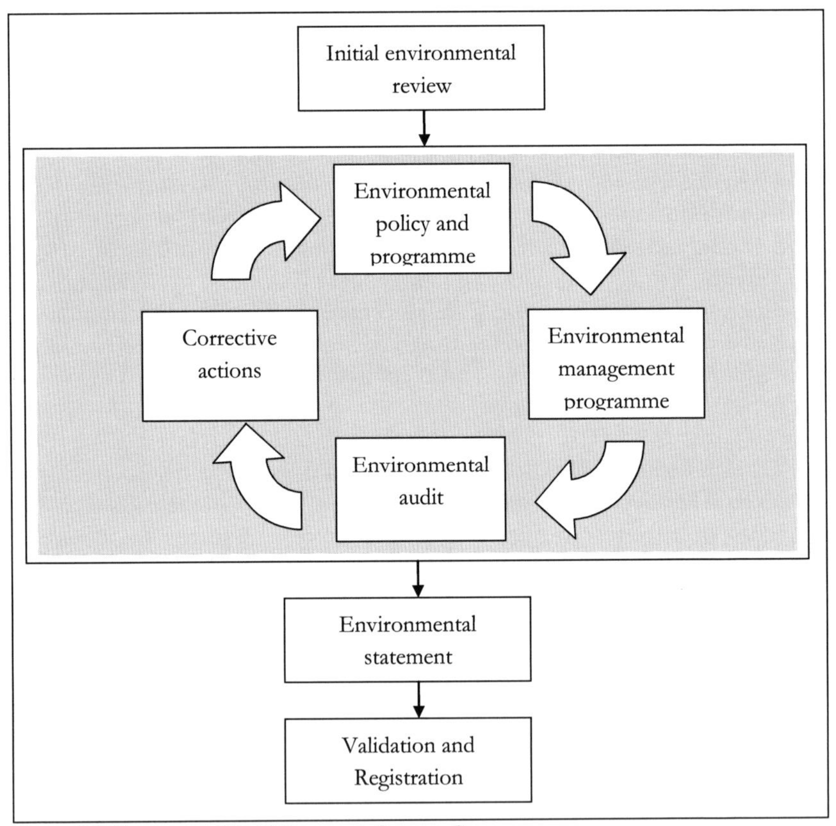

Figure 3.1: Structure of EMAS process. Source: Own.

26 The EMAS registrations' data is available through the internet via www.emas-register.eu.

27 Guidance document to the use of the EMAS logo, http://ec.europa.eu/environment/emas/pdf/guidance/guidance03_en.pdf (22 March 2006).

This diagram (figure 3.1) shows the structure of the EMAS processes, which is similar within all organisations using this environmental management system. The green box describes the continuous improvement cycle (orientated at the PDCA-cycle) that is the centre of the management system. The elements of it, together with the environmental statement and the validation and registration procedures are the main work that has to be done on a regular basis.

The EMAS regulation does not want to work like an environmental impact assessment in the sense of a complete organisational control of the environmental balance with a systematic record of all inputs and outputs, but is rather a management system with a focus on environment which, although aiming high in its targets within the regulation, sets its priorities at the following two activities: first, the organisations should go beyond the European and national environmental regulatory standards and do more than legally required. Second, the management system wants to lead to a perspective of an organisation where the improvements of the environmental outcomes are seen as a continuous process and ongoing task. Nevertheless, regarding the verifying process, the regulation is relatively weak because the auditor looks at the organisation's processes of environmental management instead of checking the real outcome of the improvements against certain standards. The focus of the EMAS regulation is on promoting ecological development in organisations in regard to the ideas, products and facilities within these organisations. It is a change within EU environmental law that traditionally has a command and control approach, thus wanting to get to a European-wide standard regardless of the individual situation within a country or in an organisation (Müller-Christ 2001).

3.3 The development of the regulation 1836/93 (EC) (EMAS I)

The history of the regulation 1836/93 (EC) (European Commission 1993), the so-called EMAS I regulation, dates back to the 1970ies when companies had to develop 'compliance audits' after several failures, especially in the chemical industry. This caused great trouble and problems because the insurance companies wanted to see compliance audits to minimise their risks when insuring large companies.[28] In the 1980ies, the U.S. Environmental Protection Agency (EPA) started several projects to implement environmental audits to promote effective and efficient environmental safety and protection at companies (Baumast 1998, p. 38-41). As early as 1988, the International Chamber of Commerce (ICC) started to standardise major procedures of an environmental management system (Delogu 1992).

A second major issue in the development of a European system for environmental management was that in the late 1980ies and the beginning of the 1990ies, the European Commission and many governments realised that they could not control the almost countless regulations of the by then command-and-control focused environmental policy, neither nationally nor European-wide. Especially the EU itself had a problem controlling its regulations because it did not have the power to do so on its own nor were the member states eager to do so for the Commission.

The hierarchical intervention, the typical instrument of European environmental policy, lost its importance at the beginning of the 1990ies when context-orientated instruments that provide more freedom for regional and local adjustments were developed, the so-called 'new environmental policy instruments' (NEPI's) - one of them being EMAS (Knill 2003, p. 63, see chapter 2 for details).

In December 1990, the European Commission DG XI (later DG Environment) started to work on a consultation paper to introduce the topic of environmental management to the European Community. The first paper, titled "A consultation paper on draft elements for a Council Directive on the Environmental Auditing of certain Industrial activities" (Würth 1993, p. 138), was based on the experiences made by the U.S. Environment Protection Agency (EPA) and the International Chamber of Commerce (ICC) as well as on knowledge gained through the use of the British BS 7750 standard. The proposal underwent several stages of change and improvement before it was passed on to the European Council in the first quarter of 1992.

In contrast to its first drafts, the final proposal suggested a compulsory introduction of EMAS for specific sectors of industry (Malek 2000, p. 68). This

---

28 One of the first companies working out an environmental audit was Allied Signal in 1976. The main aim of the audit was the collection of information of the compliance with national environmental law for the top management in order to prevent liability problems (Baumast 1998, p. 36).

text was strongly rejected by Germany especially because the federal authorities did not believe that the proposed voluntary management approach would have advantages against the traditional German compulsory technology-orientated approach.[29] As a reaction, the European Commission then proposed the voluntary use of the management system. In the course of the further development of the regulation, German industry suggested two alternatives to the Commissions proposals: first, the idea of having an environmental conservation manager (German: Betriebsbeauftragter) within every organisation who would be responsible for environmental issues.[30] A second alternative was the idea that all member states should raise their environmental standards towards the German level. Consequently, the regulation would not have been harmful for German industry. At the same time, the German government felt that the EMAS accreditation would be unfair to German companies that had to follow the (at that time) high national standards in comparison to an organisation in a country with lower standards (Franke/Wätzold 1995).

In accordance with the EU rules, the proposal was given to both the European Parliament (EP) and the Economic and Social Council (ESC) (EU Official Journal 1992). In their statements, both organisations stressed the necessity for clearly-defined, detailed standards for the evaluation of the organisational management system within the EMAS regulation. Additionally, they were of the opinion that the influence of the employees within the system was crucial to its success and that it had to be strengthened further within the Commission's proposal. In addition, the EP stressed that the regulation should not only focus on compliance but also judge the past breaches of environmental legislation and activities to prevent this in the future. The EP also wanted to increase the attractiveness of the management system by giving incentives such as tax exemptions and the possibility to give

29 This approach is rooted in the technology approach that was developed from the 1970ies onwards. Detailed technical instructions were issued to establish a certain level of environmental quality, e.g. air and water quality.

30 According to the German mode of operation, the environmental conservation manager is a member of the organisation but primarily responsible for the daily routine work of environmental issues. He is the expert for both the senior management team of an organisation and also the primary contact partner for public authorities regarding environmental issues. He is not finally responsible for environmental issues; this responsibility remains with senior management. There are also the positions of waste manager (German: Abfallbeauftragter), water conservation manager (German: Gewässerschutzbeauftragter), dangerous goods safety adviser (German: Gefahrgutbeauftragter) and other positions where applicable (Grössmann 2006).

preference to audited companies in regard to public procurement (EU Official Journal 1993).

In the first quarter of 1992, the proposal was turned down in the European Council by Germany in accordance with article 130s of the European Economic Community Treaty, the unanimity principle. In Germany, the political opinion in regard to the regulation was separated into two opponents: whereas the Federal Ministry for Environment as well as the German Bundesrat were in favour of the regulation but demanded objective criteria to judge environmental management systems within organisations (Deutscher Bunderat: Bundesratsdrucksache 222/92), the federal Ministry for the economy was totally against the regulation because it was not convinced that such a regulation was needed (Waskow 1994, p. 4). The ministry was following the ideas of the industry lobbyists, the Federation of German Industries (German: Bundesverband der deutschen Industrie (BDI)) and the Confederation of German Employers Associations (German: Bundesvereinigung der deutschen Arbeitgeberverbände (BDA)), that were not in favour of the regulation because they disliked the need to give internal information to the general public, fearing this information could also be used by rivalling companies. Additionally, they were of the opinion that due to the already high environmental standards in Germany, the industry would have to face a disadvantage in competing with other countries if the environmental management should be made compulsory. Finally, they were of the opinion that an individual, flexible environmental management system could not be standardised (Umwelt kommunale ökologische Briefe 1997, p. 7). The environmental organisations in Germany also criticised the Commission's proposal because they saw it as an instrument of public relations rather than environmental improvement (Hildebrand/Schmidt 1994, p. 69).

Despite criticism, the number of actors in favour of the regulation grew in Germany, especially because the Maastricht treaty was due to come into force in late 1993 where the unanimity principle was not necessary any more but a majority could decide on the Commission's proposal for the EMAS regulation. As a consequence, the German delegation changed its strategy from opposing the regulation to enforcement of it in order to follow its own national interests. A central argument in favour of the regulation was the fact that though being a regulation, the EMAS proposal was seen as complementary to the German law-based approach in environmental policy. Thus, the Germans ensured that the "best available technology" should be used to improve the environmental performance of industry; a principle that leads German environmental policy. However, it was weakened by the other member states by adding an idea coming from British environmental policy that the best available technology should be used "in accordance with the economic tenability".

In late 1992, the UK developed an active role in designing the regulation. It already had the well-advanced BS 7750 standard for environmental management.[31] Another active supporter of EMAS was the Netherlands, where companies already had experience of EMS standards, one of them being the BS 7750. France also supported the move towards EMAS although it kept a low profile (Bültmann/Wätzold 2000, p. 16). After the statements of EP and ESC were recognised, the European Council agreed on a text of the regulation in December 1992. Several of the Commission's proposals had already been deleted, such as the development of a European standard for environmental management by the European Committee for Standardisation (CEN) as well

31 The British BS 7750 standard for environmental management derived from a request to the British Standards Institute (BSI) in 1990. At this time, the BSI investigated the market for an environmental management standard and came to the conclusion that a number of organisations already using the quality management standard BS 5750 would be interested in using an environmental management system. Published by the BSI in April 1992 under its official title "BS 7750 Environmental Management Systems", it was designed "[...] to enable any organisation to establish an effective foundation for both sound environmental performance and participation in environmental auditing schemes" (Peglau 1996). The standard already anticipated the developments on EU level: "With a view to European developments, the new standard is currently compatible with the European Community's proposed regulation on environmental auditing" (Rothery 1993, p. 14). There are four major differences between EMAS and the BS standard. First, EMAS requires conducting an environmental review at the beginning of the implementation process that is the guideline for the process to follow whereas the initial review made compulsory by BS 7750 is not an assessable element of the management system and therefore has no consequences for the following activities. A second difference is that although both systems require defining and maintaining an environmental policy, the BS standard requires a continuous requirement for improvement, whereas the EMAS standard has a stronger formula in reducing the environmental activities as far as possible. A third difference of both systems is the requirement of the programme review. Whereas the BS standard requires an unspecific review, the EMAS standard is much more detailed and requires a regular review according to a detailed plan. The final difference between the systems is the amount and degree of publicity. Whereas EMAS requires a (under the current regulation) yearly statement or updated statement, the BS standard leaves this totally to the free will of the organisations using it.

as the idea that the European Commission should have a direct influence on the registration of the verifiers.

In June 1993, the "Council Regulation (EEC) No 1836/93 of 29 June 1993 allowing voluntary participation by companies in the industrial sector in a Community eco-management and audit scheme" was passed by the European Council (Schwaderlapp 1999). This was the first time an EMS was laid down in law. And even more interesting, it was international, i.e. European law which finally came into force in April (Große 2003). According to the regulation, all sectors of industrial commerce could take part in it. In order to fully implement the regulation into the national regulatory framework, the EU member states had to build up systems for the accreditation and observation of environmental verifiers as well as establish an infrastructure for the registration of the EMAS sites (art. 10 of the regulation).

From the industry actors both on European and on the national level, there was considerable concern as to whether it would be possible to integrate the national standards for EMS such as the BS 7750, the Irish IS 310, the French X30-200 or the Spanish UNE 77-801(2) into EMAS. In the case of the British BS 7750, the government tended to use the standard to implement the directive because it was designed to be compatible with it (Barnes 1994). Nevertheless, one of the biggest differences between these two systems was that the British national standard was company-specific whereas EMAS was site-specific (within EMAS I). Thus, a company had to implement EMAS at all sites or not use it at all, whereas with the BS standard, the company could use it at selected sites or at least at one (Barnes 1994). The issue of the different national standards and their relationship with EMAS was solved by the European Commission when it declared the national systems as being a part of EMAS or at least comparable. Consequently, the national standards became obsolete when EMAS came into force. Nevertheless, the member states kept their national EMS as these were still in use, mainly by industrial organisations. The member states could even enforce or maintain higher standards than EMAS according to article 130t of the Single European Act (SEA).

As a policy instrument, EMAS is one of just a few EU regulations of the horizontal perspective. Together with the environmental impact assessment regulation (2001/42/EC), the directive on public access to environmental information (2003/4/EC), the EMAS regulation (1836/93/EEC and 61/2001/EC) as well as the directive on integrated pollution prevention and control (IPPC directive, 96/61/EC), they form a relatively new set of overall, framing legislative norms that can be distinguished from the media-specific or issue-specific norms (Knill 2003, p. 53).

3.4 The development of the EMAS II regulation

According to article 20 of the EMAS I regulation, the European Commission had to review the regulation after five years. Therefore, in mid-1997, the revision process was started with a meeting of a commitology group of EMAS experts. After several rounds of consultation with consultants from industry, environmental groups, verifiers, trade unions and other organisations, the European Commission presented a proposal for the revision of the regulation in October 1998 (European Commission 1998, p. 5).

One proposal was the change of the relationship between EMAS and ISO 14001. The compatibility of both ISO 14001 and EMAS was largely demanded by industry because their lobbyists wanted to make both systems more attractive to companies and also intended to reduce compliance costs between the two systems whereas the political actors on the EU level hoped to raise the attractiveness for the European standards. Up to that time, an ISO certification could lead to an EMAS certificate whereas the proposal suggested making it possible vice versa so that companies already registered under EMAS would get an ISO certificate automatically. Nevertheless, the proposal did not contain further regulations on substitution that had been formulated in an earlier draft (Malek 2000).

A second change was the extension of the regulation towards other organisations, like the service sector, public authorities and farming. EMAS should now be made available

"to all organisations having environmental impacts, providing a means for them to manage these impacts and so improve their environmental performance [...]" (as finally agreed upon in the EMAS II regulation, paragraph 14).

This was a reaction of the European Commission to the developments within the member states that already had extended the use of EMAS to other organisations.

The third aspect of the revision was the change from the site-principle to an organisational principle already in place with the BS and the ISO standard.

This also implied that a clear allocation of environmental effects of individual sites was not in the focus any more. Though this is seen critically by environmentalists (Deutscher Naturschutzring 1999, Öko Institut Darmstadt 1997), the change of perspective is a definite advantage, especially for the service sector.

What was also new was the greater flexibility with the environmental declaration. The proposal encouraged all organisations to publish a statement, although this does not have to be in printed form (EMAS II regulation, article 3.3b). Nevertheless, the publication has to be validated every year instead of every three years as in the old regulation, although the yearly one can be an updated version of an older statement.

The new regulation was passed on to the European Parliament which decided in the first reading in April 1999 to ask for a larger differentiation of the organisations in regard to their environmental impacts. Furthermore, the EP demanded support for the use of the EMAS regulation for accession countries by the European Commission. Finally, it asked for an improved participation of staff in an organisation's environmental activities. Due to the fact that from May 1999, this legislative procedure was a co-decision between EP and the European Council, the Commission published a second draft in June 1999 with most of the EP's proposals which also contained the ones issued by the Social and Economic Council (SEC). This draft focused on the incorporation of ISO into EMAS and demanded a new logo to better visualise the management system. In March 2000, the European Council agreed on a Common Position towards the new proposal and stressed both the fact that EMAS and other EMS should be harmonised and that it would now be possible for all kinds of organisations to take part in it. In July 2000, the Common Position of the Council passed the EP's second reading. Despite numerous changes of the original draft by the European Commission, the EP was still of the opinion that more incentives would be needed for EMAS participants. It also demanded reasonable (i.e. reduced) registration fees. Both instruments would, according to the EP, lead to an increase in participation. And again, it stressed the importance of including the workers as well as the general public in order to create a greater awareness of the system. Finally, the European Parliament demanded better qualification of the verifiers, who needed to be qualified accountants.

The results of the second reading again led to changes of the Commission's position towards EMAS and according to article 251 EC treaty, the

Commission responded by changing its proposal at several points, thus stressing the importance of the qualification of the verifiers as well as the necessity of the member states to promote EMAS. Several other of the EP proposals were partly taken up by the Commission, others were rejected. One idea that was not taken up by the Commission, for example, was the stronger assistance for accession countries to implement EMAS in the national setting. The Commission also denied larger financial funds but offered "technical assistance for the setting up of structures to implement EMAS" (European Commission 2000/COM/2000/0512 final).[32]

According to article 252 EC treaty, a conciliation committee was needed because the European Council did not agree on all details suggested by the European Parliament. As a result of the conciliation meeting in November 2000, both institutions agreed on a final text of the regulation, stressing the importance of the voluntary approach of the scheme as well as noting the possibility that organisations which were willing could take part in EMAS. Both enforced the visibility of the system through a new logo for EMAS (Bulletin EU, http://europa.eu/bulletin/en/200011/p104037.htm, 12 April 2005). In February 2001, the European Parliament agreed in the third reading to the final text of EMAS, thus stressing that

> "[...] the future regulation, which has now been approved by both institutions, aims at broadening the scope of Regulation (EC) No 1836/93 so that the eco-management and audit scheme (EMAS) can make a bigger contribution to environmental protection. It aims at extending participation to all sectors of the economy to provide organisations which participate voluntarily with a means of continually improving their environmental performance." (Bulletin EU 1/2-2001, http://europa.eu/bulletin/en/200101/p104048.htm, 12 April 2005)

The new regulation (EC) No 761/2001 of the European Parliament and of the European Council of 19 March 2001 allowing voluntary participation by organisations in a Community eco-management and audit scheme (EMAS) was published on March 19, 2001 and thus came into force (European Commission 2001).

---

32 Although the Commission denied financial support for the implementation of EMAS within the accession countries, it financed a guiding document (DHV Environment and Infrastructure BV 2002) for these countries and other projects, like the implementation of EMAS in Romania (http://www.dir-emas.ro/eng/links.html, 07 April, 2007). See http://ec.europa.eu/environment/emas/activities/accession_en.htm (07 April 2007) for current events.

3.5 Changes from EMAS I to EMAS II

There are a number of changes from EMAS I to EMAS II, of which the most relevant will be mentioned here (Große 2005). First of all, under the new regulation, all organisations that want to improve their environmental performance can take part in the scheme whereas EMAS I was initially designed for industry. Therefore, due to the demand from organisations outside of the industrial sector, some member states, among them Germany and the UK, decided to give other organisations access to the EMS some years before the EMAS revision. With the revised regulation, the management system is now available for all kinds of organisations in all member states under NACE code; an individual national regulation has become obsolete.

Second, with the revision of the regulation, the site-focus of EMAS I has been changed to an organisational focus. Now it is possible to use EMAS for an entire organisation or parts of it. This move was welcome by the EMAS community – it was a consequence of the convergence between ISO 14001 and EMAS.

Third, in contrast to the EMAS I regulation, EMAS II now differentiates between direct and indirect environmental aspects, the later are those over "which the organisation does not have full management control or which occur at a distance" (EMAS II regulation, Annex VI, 6.3). It also takes the product aspects of the organisation into account which EMAS I did not. To make the system more visible, EMAS now has a clear and unified logo and detailed rules regarding how to use and advertise it. Yet, it still is not possible to use it in connection with products (Bayerisches Staatsministerium für Landesentwicklung und Umweltfragen 2001, p. 9).

Fourth, although the time for a revalidation remains three years, the environmental statement has to be updated and checked by a verifier on a yearly basis with the exception of small companies with less than fifty members of staff. However, the statement does not need to be published in paper any more, an electronic availability is sufficient. The yearly validation of the statement has been criticised by many actors (Clausen/Keil/Jungwirth 2002) because it seems costly and unnecessary.

These four major changes of the EMAS regulation show two antecedent developments. One is the stronger link between EMAS and ISO in order to gain ground for the European system which is weak in participant organisations against the strong international ISO 14001 norm. In contrast, the second is the wish to develop EMAS into a premium EMS above the ISO system due to the requirements for the EMAS system.

3.6 The EMAS II revision process

Similar to the EMAS I regulation, EMAS II has to be reviewed by the European Commission after five years. This process started in 2006. Before that, other activities had already prepared the way for this process.

In 2002, a large study financed by the DG Environment that specifically called on the new EMAS II regulation's first effects was published (Clausen/Keil/Jungwirth 2002). Regarding EMAS in the public sector, the study concluded that these organisations play a crucial role in promoting EMAS due to three main reasons. First, public authorities are the central actor in bringing together the various EMAS stakeholders. Second, these organisations are themselves large and have an environmental impact which can be documented and reduced with EMAS.[33] Third, public authorities can have an impact on environmentally-focused procurement through their buying power (Altmann-Schevitz et. al. 2002, p. 11). Compared with industry, the study stated that the use of EMAS in public administration was just at the beginning:

> "In many cases, EMAS is part of the organisations' overall environmental strategy, often combined with the activities to develop a Local Agenda 21." (Clausen/Keil/Jungwirth 2002, p. 38)

The main focus for public authorities with EMAS is the collection and analysis of relevant data. Despite the move towards more cost-effective management within local authorities during the last years, in many of the cases examined for the study, there were no positive financial results visible in connection with EMAS. This is due to a lack of interest in the overall financial needs of EMAS. Regarding indirect effects, public authorities often have a large impact on these, e.g. by granting planning permission, etc. Nevertheless, no large impacts have been found in this field of environmental policy (Clausen/Keil/Jungwirth 2002). Thus, it seems that compared to the potential of the policy instrument, there are relatively few activities.

---

33 For the resource consumption and potential savings in German public administration as an example see BMU/UBA 2001, p. 20.

In 2004, the European Commission issued a report for the European Council and the European Parliament to inform them about the incentives for organisations to take part in EMAS. It was based on a study among the member states and Norway on incentives and activities for EMAS on the (national) level of the member states. It concluded that although the majority of the member states with the exception of the new ones had programmes to give incentives for the management system, more activities would be necessary to make EMAS more attractive. Especially the recognition for organisations already registered under EMAS needed to be improved, as well as the attractiveness for new participants in the scheme (European Commission 2004/COM (2004)745).

To prepare the current revision process for the EMAS II regulation, the DG environment issued a study on EMAS and the Eco-label. This 'Evaluation of EMAS and Eco-label for their Revision (EVER)' aimed to provide recommendations for the revision of two voluntary instruments in EU environmental policy. Both instruments have been under constant pressure from several sides regarding their effectiveness (Kern et. al. 2001, Loew/Clausen 2005). Although highly valued by actors who use EMAS and/or the EU eco label, their level of awareness in the general public is largely criticised and leads to a low interest in continuing or beginning with both instruments. Therefore, the European-wide study named several conclusions, of which only the ones for EMAS are of interest here. First, it stated that the use of EMAS had played a significant role in stimulating environmental improvement, especially in the fields of reducing waste, water and air pollution. Nevertheless, it claimed that factors like environmental regulation or technical progress were even more important for reducing the environmental impact than having and using an EMS. Second, it found that although EMAS-registered organisations think the environmental management system is a useful tool to improve their environmental performance, there were no major quantitative studies to confirm that. Third, there was the denial of large differences between EMAS and ISO 14001 (IEFE 2005), thus questioning the relevance of EMAS in general. Regarding the indirect environmental effects which were seen as a major issue of EMAS II, the study found that the management system had had little effect on it, especially within the supply chain. Generally seen, it also found that

> "[...] the real number of companies which adopt an environmental management system, or part of such a system, due to EMAS is far higher than current figures of EMAS participants suggest." (IEFE 2005, p. 4)

In view of the question whether EMAS is an effective tool for competition, the conclusion of the study was that EMAS participants themselves think this is the case while only a minority of organisations outside the scheme agree on it.

For the revision of the regulation, the study proposed numerous targets and changes. They all went in the same direction of increasing the attractiveness of EMAS. However, the proposals vary from better support and promotion of the scheme to making EMAS mandatory and giving it a stronger product dimension (IEFE 2005, p. 10-12). For local authorities, it suggests better and more practical guidance together with the proposal to enforce the multiplier effect that public authorities have. As a result of the study, the European Commission invited a large group of national experts to a European conference in December 2006 about the future of EMAS where the DG Environment presented its ideas of the regulation's revision together with tasks to be discussed by the national experts. The Commission suggested a radical overhaul of the regulation; it wants to further strengthen the management system as the best available system on the market. One important aim is to raise the attractiveness for interested and participating organisations and finally to increase its user-friendliness and affordability. The long-term strategic target is to have 40,000 to 50,000 participants in the scheme by 2018-2020.[34] Further, with the problems of the current EMAS-regulation well understood, the European Commission is of the opinion that only an overall and general revision of the regulation is a chance for EMAS to turn it into a sizeable and good alternative to the traditional "command-and-control" legislation. Thus, it is still seen as a necessary tool in the EU environmental policy mix without any alternatives at all. In order to position EMAS as a standard of excellence, the Commission wants to reinforce the legal compliance aspect as well as performance improvements with the help of key

34 The announcement of the Commission to have up to 50,000 participants within the scheme in 2020 together with the idea of keeping EMAS as a high-class EMS was seen very critically by the majority of the national experts at the workshop in December 2006. The participants saw an unsolved dilemma for the development of EMAS because the large number of proposed participants would automatically mean a reduction of the standards of the regulation, otherwise this aim would not be possible to reach. Nevertheless, the Commission insisted on keeping EMAS as the highest available EMS standard. Therefore, it is relatively unlikely that the proposed number of participants in 2020 will be achieved in the light of recent developments of EMAS.

performance indicators that make it possible to compare the developments of different organisations.[35]

To raise the attractiveness of the system, it was suggested that organisations outside the EU would officially be registered under the scheme; under the present regulation, there is (only) a second unofficial EMAS register with the European Commission, comprising the non-EU states at their free will. Additionally, corporate and/or cluster registrations shall be made possible as well. Furthermore, the promotion of the system as well as the technical assistance shall be improved. In order to raise the attractiveness of the scheme, it will be possible to add optional reporting options, thus featuring product information or general information on sustainability. In cooperation with the member states, the Commission is planning to increase the incentives for organisations using EMAS. Further changes are to improve the user-friendliness as well as the affordability of the system through a complete re-cast of the regulation as well as to review the rules for reporting and logo-use. Finally, the rules for harmonisation, verification and the registration process will be simplified (European Commission, DG Environment/Hamon 2006).

The ideas of the Commission were commented on differently among the participants in the workshop. Whilst agreeing that it would be positive to strengthen the management scheme, the experts were surprised at the numbers of the proposed participants in the future. They saw a misconception because from their point of view and in retrospect of the last ten year's development of the scheme, it would not be possible to get so many participants and simultaneously maintain EMAS as a prime standard for environmental management, superior to ISO 14001. Overall, the majority of the participants wanted to make working with the system easier while keeping its high standard. Performance indicators, together with a set of incentives and activities for a larger public interest in EMAS were seen as needed and useful.

Participants from environmental and consumer NGO's stressed that so far, the system lacked performance requirements as well as performance indicators. Consequently, they strongly agreed on introducing these indicators into the scheme and welcomed the use of EMAS as an instrument for legal compliance but criticised that the proposed "radical change" would not take place with the suggested revision (ANEC 2006).

35 This move towards introducing key performance indicators is an international trend, e.g. in the UK, Defra is introducing these for environmental activities in line with EMS (Department for Environment, Food and Rural Affairs 2006).

Regarding public authorities, their special role with EMAS was also highlighted in the debate, although it became clear that the participants could only agree on more support for this type of organisation. A compulsory introduction of EMAS for local authorities as suggested by the European Commission was rejected and not negotiable by the participants representing local and/or public authorities.

From December 2006 to February 2007, the European Commission organised an internet-based consultation process for the EMAS revision. A total of 214 statements from 23 member states were given during the consultation process; organisations and other actors from all interested fields participated in the process. More than a quarter of all statements came from Germany, followed by Italy, the UK and Spain. 28.5 % of the participants came from EMAS-registered organisations, 15.8 % from public authorities. The overall view of EMAS was positive but the majority of the participants thought that EMAS had to be improved. Most needed were the following changes of the scheme:

- a reduction of complexity of the regulation as well as the accreditation and re-accreditation procedures
- a reduction of the costs for EMAS
- an improved process of registration
- an increase of the external effects that can be achieve with EMAS and finally
- a larger recognition of the achievements that are possible with EMAS among the general public.

Most participants expected EMAS to continue as a voluntary system, although about half of them said that they could agree to make it mandatory for specific sectors of industry, e.g. for organisations that get EU founding, for sectors with high environmental risks or for large public authorities. The majority of participants also agreed on consolidating EMAS in order to maintain its status as the best available EMS on the market. This status should further be maintained through improvements within legal compliance and through improvements on the organisation's capabilities. A large part of the consultation covered the area of improving the attractiveness of EMAS; the majority of participants of the consultation was in favour of a stronger integration of the system into national law, an increase in the national authority's activities to consider incentives for EMAS as well as professional campaigning for EMAS (EMAS Helpdesk 2007).

According to information by the German responsible body UGA, the Commission had planned to officially present a first draft of a revised regulation in the first quarter of 2008. It was finally published unofficially on 16 July 2008 on the European Commission's EMAS homepage. An official draft publication is still due (as of December 2008). As suggested above, the European Commission wants to increase both the attractiveness of the policy instrument as well as the number of users. The major changes are as follows: a stronger focus on legal compliance - non-compliance and sanctions following it are defined in detail, a harmonisation of the rules for accreditation and validation, the official registration of non-EU organisations and an increase of incentives to make EMAS more attractive. Furthermore, the European Commission suggests the publication of good management practices documents in order to support organisations interested in using the scheme (European Commission, DG Environment 2008, Umweltgutachterausschuss 2008). The current plan is that the new EMAS regulation will come into force in 2010.

3.7 Participants in EMAS and ISO

Although EMAS has incorporated ISO with the EMAS II regulation, it is regarded as the qualitatively higher standard due to the more detailed demands. While EMAS is based on an EU regulation, governed by the Commission and implemented by the member states, ISO was created by an international industrial association and national bodies of standardisation. Whereas EMAS requires the improvement of environmental performance, ISO asks for the improvement of the environmental management system. Regarding the relation to environmental law, EMAS organisations must assure their regulatory compliance while organisations using ISO are required to commit themselves to this compliance. Additionally, whilst EMAS requires a public environmental statement to be published regularly that has to be checked by an independent verifier, there is no such publication necessary for ISO 14001 users. To conclude: EMAS is considered to be the standard of environmental excellence; ISO is a method for the standardisation of an organisation's environmental activities (Bracke/Albrecht 2007).

Regarding the central elements of both EMS, there are considerable differences between EMAS and ISO 14001:

	EMAS (I and II)	ISO 14001
Initial environmental review	Compulsory, relevant for the evaluation of the environ-mental impacts of an organ-isation	Voluntary or optional
Environmental policy	Necessary for legal compliance	Requirement to commit to compliance
Environmental targets and environmental programme	All environmental impacts have to be assessed, targets have to be quantified, time frame is necessary	Planning of environmental activities necessary, no exact time frame needed
Environmental management system	Site specific (EMAS I) Related to organisation (EMAS II)	Related to organisation
Environmental audit	Carried out by independent verifier, at least every three years	Related to the system itself, no detailed audit time frame, internal management review
Environmental declaration	Compulsory, content according to EMAS regulation	Not applicable

Table 3.1: Differences between EMAS and ISO. Source: Own.

Both systems follow a common idea but have a different legal basis and differ in numerous elements. The number of participants in EMAS is considerably lower compared to ISO 14001. As of July 2007, the EMS standards ISO 14001 and EMAS are used as follows:

	ISO 14001	EMAS
Worldwide	129,031	5,389
Germany	5,800	1,979
United Kingdom	5,400	364

Table 3.2: ISO certifications and EMAS registrations, including non-EU EMAS users, as of December 2006/January 2007. Source: Peglau-Liste http://www.umweltbundesamt.de/umweltoekonomie/ums-welt.htm (20 March 2008)

These figures clearly indicate that the ISO standard is by far the more successful. Nevertheless, one can also see that EMAS is much more widespread in Germany compared to the UK where the difference is extreme, thus showing that quantitatively EMAS is not successful here.

The EU regulation is the framework for the EMAS implementation within each country. While the general structures should be the same in Germany and the UK, Bültmann/Wätzold (2000) come to the conclusion that from a company's perspective, there are hardly any advantages of EMAS over ISO 14001 in the UK, thus explaining the very low number of participants in EMAS. In Germany, the situation is slightly different. EMAS I had an advantage in the case of regulatory relief over ISO 14001. Regarding founding and external communication, EMAS is also seen as advantageous for companies over ISO 14001.

	UK	**Germany**
Participation costs	ISO 14001	ISO 14001
International recognition	ISO 14001	ISO 14001
Clarity of EMS	ISO 14001	ISO 14001
Similarity to ISO 9000	ISO 14001	ISO 14001
Regulatory Relief	Equivalent	EMAS
Involvement of Business organisations	Equivalent	Equivalent
Promotion (information)	Equivalent	Earlier: EMAS, now: Equivalent
Promotion (founding)	Equivalent	EMAS
External communication	Varies according to company	EMAS

Table 3.3: Advantages of EMAS and ISO 14001 from companies' perspective. Source: Bültmann/Wätzold 2000, p. 40.

There are numerous studies to explain these national differences. While Perkins and Neumayer (2004) suggest that EMAS registrations are higher in those EU member states with a less interventionist and burdensome style of regulation, these findings do not explain the relatively high number of EMAS participants both in Germany and Austria because these countries have a particularly high regulatory burden. Delmas (2002) has found that the commitment of a government to environmental goals has a positive influence on the uptake of ISO 14001 by companies. Further, Potoski/Prakash (2004) made evident that the rates of take up ISO are likely to be higher in countries with governments that are flexible with the stringent use of environmental regulations. Overall, it is the support of the industrial organisations, their lobbying organisations and the knowledge, usefulness and esteem of the system which seem dominant factors for the high uptake of ISO 14001 and the relatively low uptake of EMAS.

The overall situation with EMAS is presently (December 2008) that EMAS has more than 6693 registered sites and 4138 registered organisations which are the highest numbers since the regulation was published (EMAS Helpdesk 2008). Currently, especially Italy, Spain and Portugal are strong with rising numbers of registrations while Germany, the leading nation since the beginning, is currently losing organisations taking part in the scheme. Overall, about sixty per cent of the participants in the scheme are SMEs; the main sectors of operation are chemicals and food processing (European Commission, DG Environment/Hamon 2006, p. 8). The increasing numbers of EMAS participants is due to a rising interest in environmental issues, more legal incentives as well as more political and financial support from national and sub-national authorities in Italy, Spain and Portugal.[36]

3.8 The implementation of the EMAS in the UK

In 1992, the British Standards Institution (BSI) introduced the BS 7750 standard for environmental management. It was developed primarily because British industry had called for an EMS standard due to a surge of environmental concerns in the 1980ies and early 1990ies. The standard, therefore, was developed in close cooperation with several industrial organisations (Delgado 2000, p. 68). The BS standard was the predecessor and model for the ISO 14001 standard that was introduced in 1996. In 1996, the BS standard was recognised being compatible to EMAS (European Commission 1996a).

For central government, there was no need to integrate the EMAS regulation into the UK legal system because it came automatically into force (Wölk 2002, p. 199). Consequently, there is no associated legal regulation associated with EMAS in the UK. Nevertheless, the government had to make various

36 In Italy, for example, EMAS is combined with IPPC so that organisations having EMAS get credit for the on-site checks for the IPPC regulation. In Portugal, there is an unwritten agreement on the national level between the EMAS responsible body and the Inspectorate for the environment whereby EMAS-registered organisations are less controlled by the Inspectorate, assuming that these organisations have better control systems of their environmental impacts. In Spain, the sub-national authorities have set up a number of initiatives and structures to strengthen the implementation of EMAS, e.g. the region of Galicia is covering up to 75 per cent of the costs of registration and verification. The national government has published a guide for SMEs to assist with the implementation of EMAS.http://ec.europa.eu/environment/emas/activities/index_en.htm (8 April 2007).

arrangements to ensure the working conditions of the regulation by organising or naming a responsible body and an accreditation authority. The UK government was the first of the EU members to participate fully in the scheme (Strachan et. al. 1997) with the establishment of the institutions needed. This was due to the prior experiences with the national BS standard for environmental management. From 1996 until 1998, registration of EMAS organisations was done by the Ministry for the Environment which became the competent body according to article 3 of the EMAS I regulation. Then, the ministry gave this role to the Institute of Environmental Assessment (IEA), a non-profit organisation founded in 1990 working in the field of environmental management. In 1999, the IEA together with the Institute of Environmental Management and the Environmental Auditors Registration Association merged into the Institute for Environmental Management and Auditing (IEMA). It has been the UK's competent body for EMAS since then (http://www.iema.net/iema, 05 March 2007). The responsibility for the accreditation of the verifiers has been allocated to the Department of Trade and Industry which has given it to the quango 'National Accreditation Council for Certification Bodies' (NACCB). In 1995, this organisation was renamed 'UK Accreditation Service' (UKAS). It is comparable to the German accreditation body for environmental verifiers (Kähler/Rotheroe 1999, McIntosh/ Smith 2000) and responsible for the definition of standards and competences of the verifiers of all three EMS systems BS 7750, EMAS and ISO 14001. It is a publicly limited company with members representing central government, British industry and industrial standards organisations. Furthermore, UKAS is responsible for a number of other accreditation regulations regarding laboratories etc.

Overall, the implementation process and the creation of the accreditation, supervision and registration system did not lead to major conflicts or large discussions, like in Germany (Bültmann/Wätzold 2000a, p. 36). Although the necessary organisational structures were implemented before EMAS came into force in May 1995 because the national BS standard was already in use, by that time, the government regarded the establishment of the BS 7750 as an advantage for UK companies. The idea (of central government) that the British standard would be a step towards EMAS was rejected during the development of the EMAS regulation on EU level by Germany, who was of the opinion that national standards should only be recognised after the agreement to an international standard.

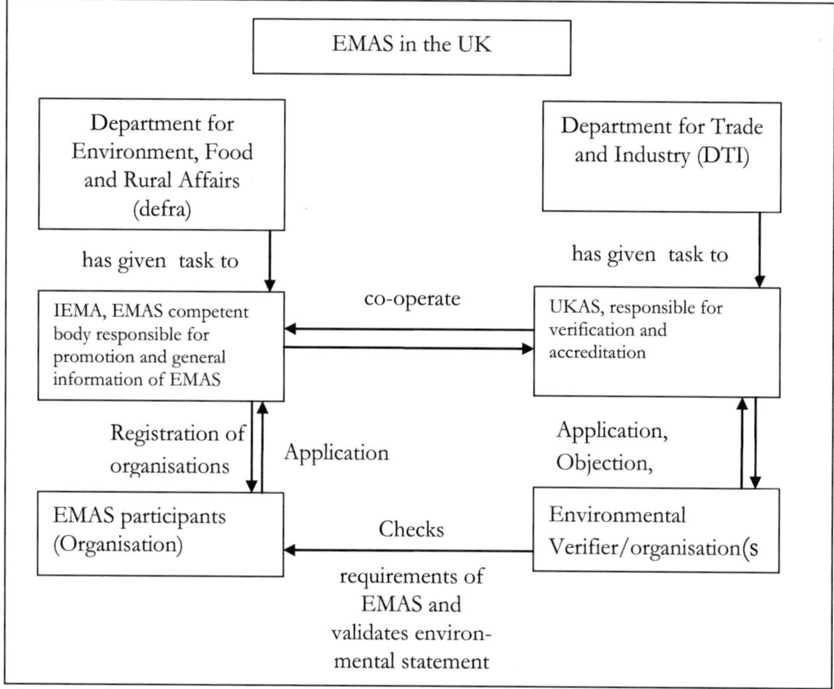

Figure 3.2: Structure of organisations responsible for EMAS in the UK. Source: Own.

Consequently, it was not until February 1996 that the BS 7750 standard was recognised as meeting the requirements for EMAS. Due to the fact that, at the same time, the internationally recognised ISO 14001 standard was to be put into force, the BS standard became obsolete and so did the idea of the British government that the national standard would be a "door opener" for the European one (Bültmann/Wätzold 2000, p. 35-36).

Regarding UK legislation, there are no formal links between EMAS and the framework of UK mandatory environmental regulations. Thus, the incentives for organisations to participate are relatively low. One major issue is the UK's climate change levy, introduced in April 2001, which aims to drastically reduce the output of carbon dioxide by industry. Companies with an energy reduction plan can reduce the levy by eighty per cent. EMAS can be used as a tool for verifying this plan (EMAS Help desk, http://ec.europa.eu/environment/emas/activities/index_en.htm#uk, 24 February 2007). A second issue is the connection with the risk rating scheme under the EU's 'Integrated Pollution Prevention and Control regulation' (IPPC): the UK

Environment Agency gives the highest recognition to EMAS, more than to ISO 14001. In connection with the UK's Operator and Pollution Risk Appraisal (OPRA) scheme, this can be used to decrease the charges for the relevant permits (European Commission 2004, p. 6). Furthermore, the BSI published the standard BS 8555:2003 in 2003 which is a guide to the implementation of environmental management systems. It can be used as a help to introduce EMAS but there is no incentive connected with it. There are no incentives on public procurement nor is there any financial support (European Commission 2004, p. 18).

3.9 The development of EMAS in the UK

As the BS standard was developed and introduced before EMAS, and the ISO 14001 standard was seen as directly developed from the BS standard, it was easy for a large number of organisations to implement the ISO environmental management system after already being registered under BS. On the contrary, the move towards EMAS for organisations already using BS or ISO was easier compared to those who had no EMS at all (Kähler/Rotheroe 1999). Nevertheless, the number of EMAS participants in the UK developed on a very low level compared to a large number of EU countries, not to mention in comparison to Germany. It has become stable on a very low level since 2002.

Generally, the introduction of EMAS in the UK was supported by the Department for Environment (DfE, now Defra[37]) with a programme for SMEs. Under this scheme, more than fifty companies received half of the costs for external consultants to introduce EMAS. Additionally, the DfE started a campaign to promote EMAS among industry (Department of the Environment 1997, p. 76). This campaign was terminated in July 1997 due to low participation rates. No other programme to support EMAS for any type of organisation has been organised after that by national government.

In comparison to other EU countries, the idea to introduce EMAS in public authorities was taken up as early as 1993 (Jacobs/Levett 1993).

---

37 Defra was formed in June 2001 when the Ministry of Agriculture, Fisheries and Food (MAFF) together with the Department of Environment, Transport and the Regions (DETR) and a unit of the Home Office were merged into a new organisation as part of the reaction caused by the foot and mouth disease.

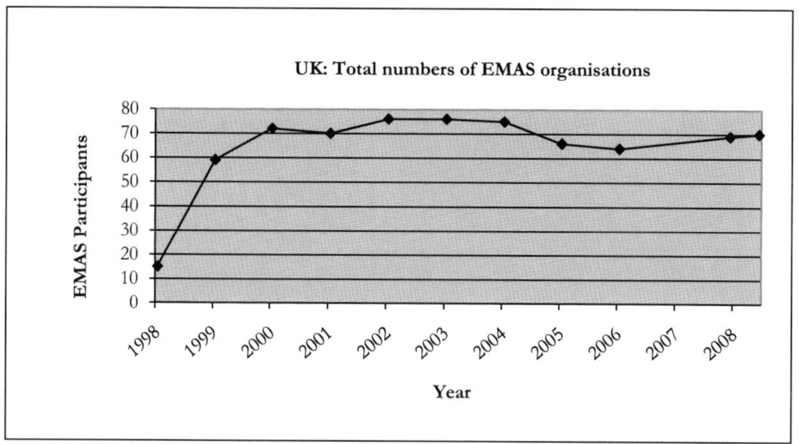

Figure 3.3: Total number of organisations participating in EMAS in the UK. Data for 2007 as of December 2007. Data for 2008 as of June 2008. Source: EMAS Help desk.

Supported by the central government, the DfE together with the Scottish Office worked on a programme to implement EMAS in local authorities. Together with the Local Government Management Board (LGMB), an organi-sation that supported the modernisation of local public authorities and was financed by the Office of the Vice Prime Minister (OVPM), the DfE established an information office in 1995 to promote EMAS in public authorities. In 1996 and 1997, the interest in EMAS rose and by the end of 1997, about thirty-three organisations were working on an environmental management system and three local authorities finally registered under EMAS. The number of sites (an important criterion under EMAS I) rose to nineteen, but seventeen of these were part of the London Borough of Sutton. At the same time, 113 local authorities were considering the introduction of EMAS. The outcome of the project was an adapted regulation that was developed into a UK circular (Department of Environment Circular 2/95 which formally set out requirements for EMAS in Local authorities (LA-EMAS)). Equivalent regulations have been published by the Welsh Office and the Scottish Office.

According to the DoE circular, EMAS for local public authorities differed from its use in industry in the following three points: first, the organisational approach was taken up by local authorities, referring to "operational units" rather than using the "site"-focus of industry (as introduced in the EMAS II regulation for all organisations taking part in the scheme). Second, under this organisational view, an organisation could decide whether to register as a whole, in parts or only with a unit. This was not possible for industry under the site-viewpoint. Third, within local authorities, less emphasis was given to direct aspects; thus, they should focus more on the impacts that occur as a result of the services provided (DoE Circular /University of West England, http://euronet.uwe.ac.uk/emas/brief.htm). With the adoption of EMAS II, all these aspects were incorporated in the new regulation and gave more room for the particularities of local authorities. With the revision of EMAS II, it seems likely that these aspects will be considered even more. Thus, the DoE ideas on EMAS within public authorities were very advanced and have proved to be right – but not through the numbers of participating organisations within the UK. Politically, the positive trend towards EMAS within local authorities was due to several reasons: the introduction and use of external review in general for public authorities, rising importance of environmental questions among the British public as well as an increasing number of full-time employees who work on EMAS (Riglar 1997). In the following years, the interest in EMAS rose slightly until 2001 when it began to drop again.

In the UK, the only activities to promote EMAS in public authorities were the ones initiated by the LGMB between 1995 and 1997 (Riglar 1995, Riglar 1997). Despite the overall low participation in the scheme, there are several authorities that have shown an exceptional commitment to EMAS, among them Kirklees Metropolitan Council and Lewes District Council. Both are very active with their environmental management and promote their EMAS activities not only on a national but also on an international level.[38]

To conclude, over the years, the overall number of participants rose to seventy-six organisations in 2001 and has remained on this constantly low level. Up to the present, central government promotes BS 7750 and ISO 14001 as complimentary. It did not take the position of actively promoting EMAS as a superior or more preferable EMS. Although there has been a debate to award deregulation to EMAS-registered authorities, there have only been minor activities.

One can see that when EMAS was developed on the European level, it was in line with the UK's environmental policy and thus the implementation was quick, effective and with regard to the special needs of local public authorities

38 For Kirklees see e.g.: http://www.ec.europa.eu/environment/emas/toolkit/toolkit_4_3_11.htm, 23 February 2007, for Lewes see e.g.:http://ec.europa.eu/environment/emas/local/pdf/ la_toolkit_commission_020204_en.pdf, 23 February 2007.

Figure 3.4: EMAS-registered public authority sites in the UK under NACE 75 as of February 2008. Red: local authorities. Green: central government authorities. Source: www.emas.org.uk, 27 February 2008.

and the following revisions of EMAS I (and the forthcoming proposed revision of EMAS II), quite progressive. However, the overall development of EMAS has made it clear that the regulation is not widespread within the UK. This is due to a low profile of the management system – it is known only by few experts. Additionally, there is no support from central government for it. Furthermore, and most importantly with a view to local authorities, it is not integrated into the UK's environmental policy regarding the tasks of local public authorities nor is there an integration of EMAS into strategies of performance measurement and auditing. Because these performance standards are the guideline for operation for local public authorities and determine their activities, there are only few incentives for local authorities to use EMAS. It is primarily the interest and environmental awareness of single authorities or – to be more precise – of leading actors within these authorities that promote the introduction and implementation of a European environmental management standard. Therefore, the same question as in other EU countries (among them Germany, see below) remains: how will EMAS develop in the future? The declining trend of participation is ongoing. This might change when the EU regulation is revised and has come into force in 2010. Certainly, incentives, an active marketing and the integration of EMAS into other policies are necessary to raise the numbers of participants. However, more needs to be done to keep EMAS alive in the UK.

3.10 The implementation of EMAS in Germany

For the German environmental policy, the EMAS regulation is a new instrument of indirect control within environmental policy (Héritier et. al. 1994, Jänicke/Kunig/Stitzel 1999, p. 211, Knill 2003, p. 62). Up to the development of the EMAS regulation, the German approach to improve environmental outcome has been "engineer-driven", meaning that typically, German companies tried to develop or install new technology to be more environmentally friendly. The German environmental policy traditionally reflects this approach with its detailed technical regulations called "Technische Anleitungen" which set the norms for engineered technology to reduce its impact on the environment. According to Bültmann/Wätzold (2000), the management-orientated approach that is behind EMAS was "alien to German companies". Nevertheless, it was implemented into German law as a result of European integration. This process and its specifics will be described in the following.

According to EMAS I, the national responsible authorities had the task to implement the community law into national law and establish the accreditation authority and the responsible body as the two relevant organisations according to article 18 of the regulation. Due to the fact that the regulation was a new approach in comparison to the command-and-control approach of German environmental law, there was a long discussion about the design of the responsible bodies[39] between the Federal Ministry for Environment (German: Bundesministerium für Umwelt, Verbraucherschutz und Reaktorsicherheit, BMU) and the leading organisations of German industry, headed by the Federation of German Industries (German: Bundesverband der deutschen Industrie, BDI). Each of the parties developed a model for the design and control of the relevant authorities, favouring their own position. These two models differed mainly in the institutional structure of the responsible body and the accreditation authority.

As early as 1993, the German Federal Ministry for the Environment (BMU) together with environmental groups favoured a model whereby the certification of knowledge of the verifiers would be done by a state-controlled organisation under participation of industry. The control and accreditation of the verifiers was to be done by the Federal Environment Agency (Umweltbundesamt, UBA). Furthermore, a committee that consisted of members of federal and state authorities, industrial and environmental organisations as well as labour unions were to be established which would work out recommendations and guidelines, and give advice to the federal government. The relevant authority of registration should be named by the Länder, not by the federal government (Malek 2000). In contrast to this, the industrial organisations headed by the BDI preferred a model whereby most of the control and initiative would stay with industry. The idea was a model that gave the chambers of commerce a central role in the system because they would have been responsible for the accreditation and control of the verifiers as well as the registration of EMAS organisations.

In late 1994, Federal Government and the BDI took up talks to come to an agreement. Together with the Länder, environmental organisations and the trade unions, the central actors reached a compromise in early 1995 which comprised central issues of both models (see figure 3.5). The BMU's position is reflected by the structure of the 'Umweltgutachter-ausschuss' (Environmental Verification Committee), which has the responsibility to develop guidelines for examination and accreditation and is responsible for the promotion and further development of EMAS in Germany under the legal control of BMU. It consists of twenty-five members of the following groups: industry

39 Malek 2000, p. 162, notes that the voluntary management approach could on the long run be a starting point to modernise German environmental law. This note made in 2000 has not become true so far.

and enterprises (6), environmental verifiers (4), federal or state governments in environment (6) and in economics (3), trade unions (3) and environmental NGOs (3). The position of the BDI and the other industrial organisations is reflected by the creation of the accreditation body for environmental verifiers (German: Deutsche Akkreditierungs- und Zulassungsgesellschaft für Umweltgutachter mbH, DAU), which is an industry-led organisation, as well as by the function of the chambers of commerce, which are the registrars for the EMAS organisations.

This solution secured the positions of all actors of the EMAS policy. For Federal Government, it was important to have reached this agreement because otherwise industry's support for the new policy instrument would have been endangered. The compromise agreement had to be passed by the Bundestag and the Bundesrat, and finally came into effect in December 1995.[40] The German environmental audit law (Umweltauditgesetz, UAG)[41] reflects the interests of industry[42] and the other stakeholders, and covers the accreditation and supervision of the verifiers through the industry-financed agency DAU. This organisation is responsible for the setting of general regulations for the verifiers. The actual accreditation of the verifiers is done in cooperation between DAU and UGA: both are under the legal control of the BMU.

The UGA members come from the following institutions or sectors: industry, verifiers, federal and state administrations, environmental organisations and trade unions. Through the cooperation of all actors of society – which is a novelty in German environmental law (but not a novelty in German policy

---

40 According to the EU rules for national implementation of the regulation, it had to be implemented by May 1995. The environmental ministers of the Länder agreed on a preliminary solution to establish a committee that would have the same actors as the later Environmental Verifier Committee (German: Umweltgutachter-ausschuss, UGA) because the overall solution to the discussions mentioned above had not been found yet.

41 Although the EU law on EMAS was a regulation and, thus, a final regulatory action, the German 'Umweltauditgesetz' was necessary to establish and implement the relevant bodies for EMAS (Bundesministerium der Justiz 1995, Wölk 2002), latest version of German UAG http://bundesrecht.juris.de/uag/BJNR159100995.html, 22 March 2008).

42 This is due to the fact that EMAS is a voluntary instrument that was designed for industry in the first place. Therefore, it would not have made any sense to totally oppose the ideas of industry because otherwise no organisation would have implemented it.

design, which often reflects on groups of civil society) – one tries to achieve a great acceptance of the EMAS process as well as getting the knowledge that is necessary for the sometimes complex environmental issues. Consequently, the UGA is the centre of the EMAS policy community in Germany. It brings together all relevant actors and, additionally, keeps contacts with the interested public, e.g. the scientific community.[43] The registration of the EMAS-verified organisations is done by the regional chambers of industry and commerce in cooperation with the verifiers.

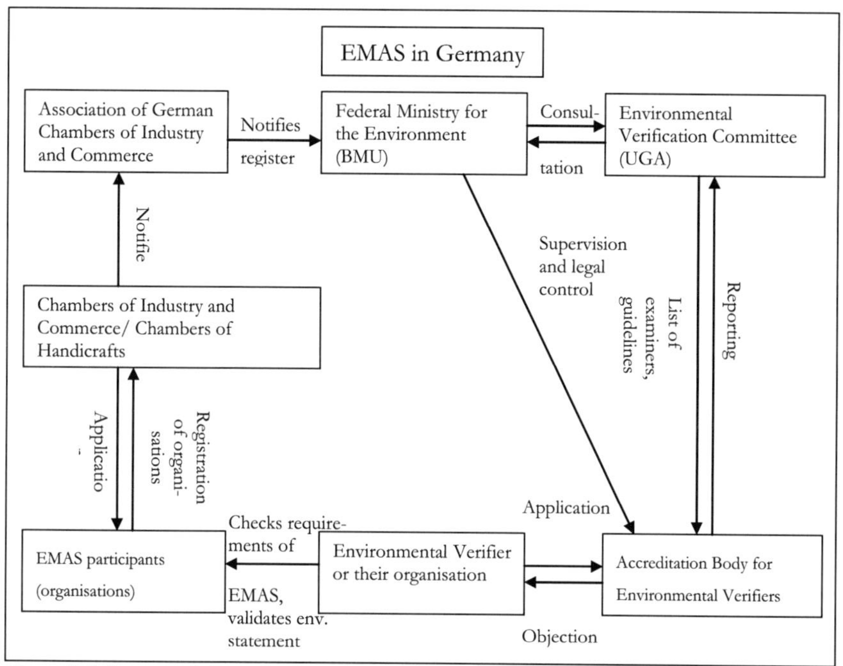

Figure 3.5: System of Accreditation, Supervision and Registration according to EMAS II in Germany. Source: http://www.uga.de/downloads/Orga-EMAS-eng.pdf, (8 April 2007)

The EMAS I regulation was focused on industry but soon after the introduction of the regulation, there was a great demand in Germany from other organisations to take part in the scheme. The first projects began as early as 1997. Even before the regulatory framework was set in Germany (Hermann/Borguslawski 1997), the first local public authority was registered under EMAS in December 1998. Consequently, the German regulation on

43 Italy, Austria, France and the Czech Republic have EMAS committees of a similar style (UGA 2005, p. 15).

extension of the EMAS participation (German: UAG-Erweiterungs-verordnung, UAG-ErwV) in 1998 made way for service organisations, banks, insurance companies, trading companies and public authorities to use the EMS.[44] It was the predecessor to the EMAS II regulation which opened the management system for all organisations under the NACE code.[45]

3.11 The development of EMAS in Germany

Although German industry was not in favour of EMAS due to a lack of concrete standards and the fear that the requirements for validation would differ in the EU member states and thus would lead to a disadvantage for the German industry, (Héritier et. al. 1994 p. 298), the development of EMAS in Germany was very positive until 2001. Nevertheless, there were rising concerns about the effectiveness of the implementation (Bund-/Länder-Arbeitskreis steuerliche und wirtschaftliche Fragen des Umweltschutzes, without year, Umweltbundesamt 1998, Kähler/Rotheroe 1999, p. 123, Loew/ Clausen 2005).

A good indicator for the EMAS discussion in Germany, especially in the public sector, is the monthly magazine 'Umwelt kommunale ökologische Briefe' (from 2008: 'UmweltBriefe', two issues a month) made for expert staff of local public authorities working in the field of environmental policy. From 1997 until 2000, mainly positive news and current issues about EMAS were discussed (eg. Hermann/Boguslawski 1997, Lütkes 1999) and several organisations, primarily public authorities, described their developments while introducing an EMS (e.g. Umwelt kommunale öffentliche Briefe 2001). These articles reflected different viewpoints towards EMAS and described support instruments at the sub-national Länder level.

---

44 Even before the German extension of EMAS participation, some organisations implemented EMAS. One of the first was a local authority, the environmental department of the city of Hanover. See Landeshauptstadt Hannover 1998 and 1999 for details. The regulation stated that not all public authorities could take part in EMAS but only local authorities. With EMAS II, this rule became obsolete.

45 The NACE code represents a European statistical classification of organisations according to their economic activity. A full NACE code list can be found in Council Regulation No 3037/90 of 09/10/1990, OJ L 293 of 24 October 1990, as amended by the Commission Regulation (EEC) No 761/93 of 24 March 1993 and the Commission Regulation (EEC) No 29/2002.

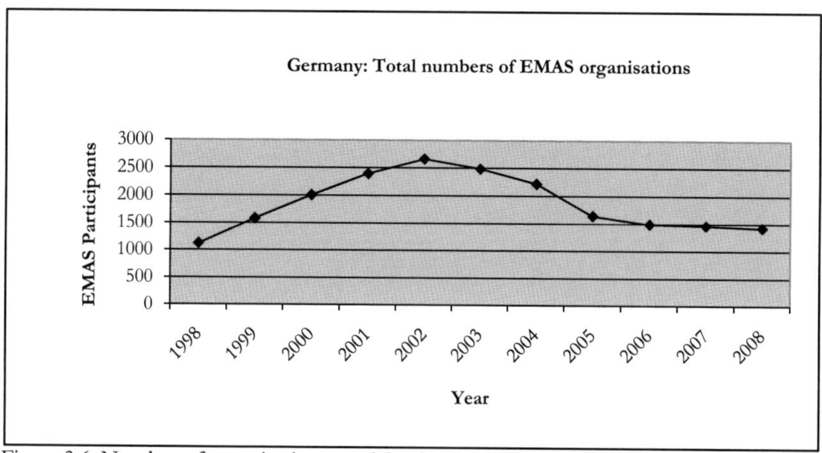

Figure 3.6: Number of organisations participating in EMAS in Germany. Data for 2007 as of December 2007. Data for 2008 as of June 2008. Source: EMAS Help desk.

In addition, they described the development of EMAS within single projects. From 2002, the number of articles published about EMAS decreased significantly, thus reflecting an overall reduction of interest in the management system. This was paralleled in the number of participating organisations in the scheme. Articles changed from individual project stories to a more general discussion about EMAS itself, its usefulness and problems (Clausen/Jungwirth 2002).

Looking at EMAS in general, from 2002 onwards, the numbers of organisations using EMAS has decreased due to a lack of interest in the system and in environmental issues as a whole although one can assume that the real number of organisations that have an EMS or parts of EMAS is higher than the number of participants suggest (Loew/Clausen 2005, IEFE 2005, p. 4).

From the very first days of the implementation of EMAS, one major interest for many organisations to use the scheme was deregulation because there are about 8,000 environmental laws, regulations and ordinances coming from European, national or regional level (Schmitz 1997). Due to the federal structure of Germany, the Länder are responsible for environmental law and can use deregulation as a means to influence policy. Regarding EMAS, these deregulations have taken a very long time and are still not very common although there are a number of instruments used to for deregulation (Umweltgutachterausschuss 2006):

- environmental alliances (German: Umweltallianz, Umweltpakt, Umweltdialog, Umweltpartnerschaft or similar) between federal state, industrial organisations/chambers of commerce and sometimes NGOs on a voluntary basis
- Industry commits itself to being environmentally active beyond law and gets different forms of deregulation like financial aid, fewer state controls, fewer rules to follow etc.
- exemption from complying with the law
- reduction of fees according to federal emissions law (Bundesimmissionsschutzgesetz, BImschG) and other federal laws
- deregulation in water law.

Since the very beginning of the discussion to support participants of EMAS through deregulation, organisations using the management system have claimed that not enough has been done to keep the system attractive through deregulation (BLAK, without year, Clausen/Keil/Jungwirth 2002, Loew/ Clausen 2005). UGA's collection of information (UGA 2006) reflects this position because only a few Länder have fixed standards of deregulation; the majority of them only give deregulatory relief on request and/or by voluntary agreement of the relevant body. This has been the case for some years now (Bültmann/Wätzold 2000, p. 50-51). According to UBA (UBA 1999), the central reasons for participating in EMAS were the idea of the continuous improvement of the organisational environmental aspects, followed by the identification of weaknesses regarding the use of energy and resources. Other reasons were the motivation of staff, the improved image of the organisation and the increased legal certainty.

Regarding EMAS in public administration, the German focus was to implement EMAS in local authorities. This was done on the sub-national Länder level. Especially Baden-Wuerttemberg and Bavaria promoted EMAS to local authorities to a large extent. The first supported the implementation of EMAS through convoy projects in which a group of local authorities worked together to achieve the EMAS validation for each organisation. From 2000 until 2006, Baden-Wuerttemberg spent 1.2 million Euros on these projects, also including participants from different sectors of industry as well as public authorities (Landtag von Baden-Wuerttemberg 2006). Around one hundred public authorities, the majority of them being local authorities (Gemeinden, Städte and Landkreise), have taken part in EMAS projects. Numerous publications, conferences and meetings have taken place (Landesanstalt für Umweltschutz 1998, Landesanstalt für Umweltschutz 2002,

Landtag von Baden-Wuerttemberg 2006a). One of the most significant was a national conference on EMAS in public authorities in Ulm in 2003 (LfU Baden-Württemberg/LfU Bayern 2003). Currently, a working group is developing a new strategy for EMAS (as of April 2007, the state agency for environmental affairs, LUBW Baden-Württemberg 2007).

The Länder Bavaria and Schleswig-Holstein have also done considerable promotion for EMAS. In Bavaria, the state government promoted the use of EMAS through numerous incentives, publications (Bayerisches Landesamt für Umweltschutz, without year) and a pilot programme from 1998 to 2000, encouraging towns and cities to participate in EMAS. This has lead to a considerable number of local public authorities using the scheme (BLAK, without year). The state of Schleswig-Holstein was also one of a few that supported EMAS in local public authorities. One project was aimed at combining EMAS and public authority reform under NPM (Grahl/Falkenberg 2001).

The activities of the Länder have had a considerable impact on the geographical allocation of EMAS in Germany. There is a direct link between the number of participants and the activities and support of each state, as shown in figure 3.7, which indicates the participants registered under NACE 75.1. Additionally, there are numerous other public authorities running an EMS under EMAS beyond the initiatives of the Länder. From 2002 to 2004, the Federal Environment Agency (UBA) initiated a pilot project to implement EMAS in federal authorities (Umweltbundesamt http://www.umweltbundesamt.de/EMAS/forum/) which ended with the registration of three of the four organisations taking part. Numerous other public authorities, like hospitals, fire stations, universities, waste management plants etc. have also implemented EMAS over the years (Clausen/Keil/Jungwirth 2002).

To conclude, one can say that the introduction of EMAS has been a long-lasting and complex process but it resembles the German federal corporatist state, especially regarding the participation of the different interest groups in the system as well as the promotion of the management system by the Länder. Thus, the organisational structures of EMAS were one central factor for the success of the EMAS regulation up to 2001.

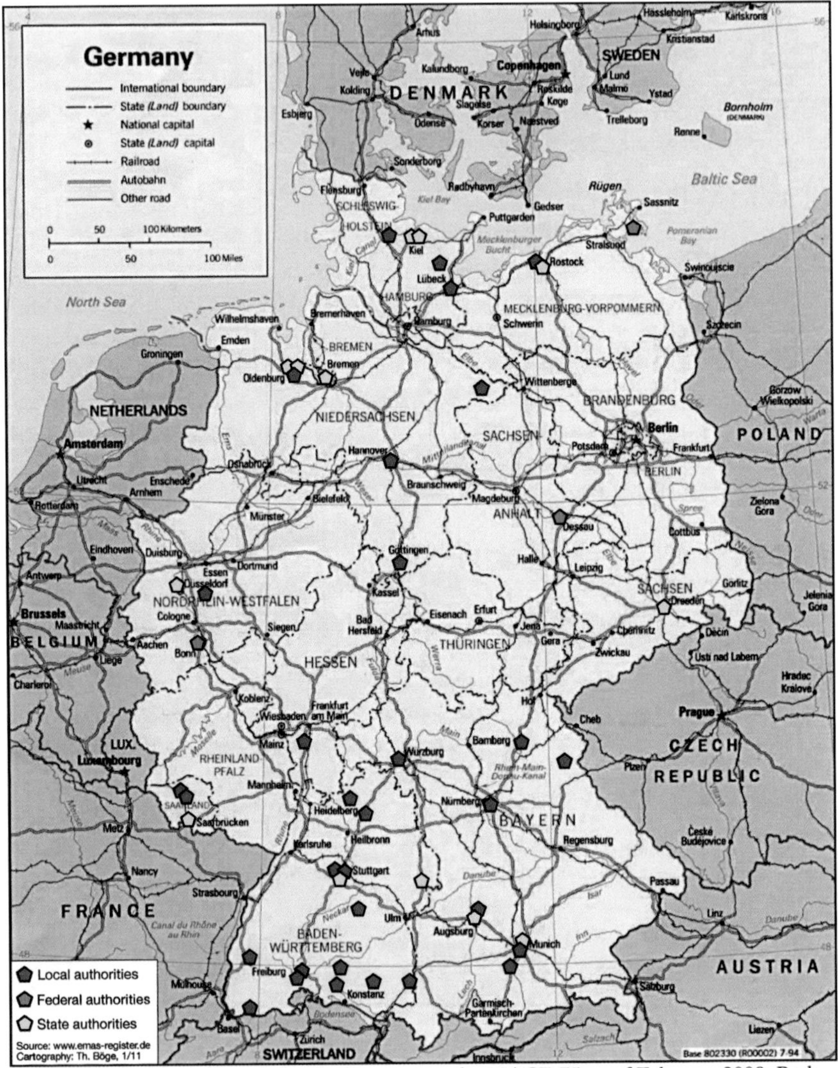

Figure 3.7: EMAS-registered sites in Germany under NACE 75 as of February 2008. Red: local authorities, Yellow: state authorities, Green: federal authorities. Source: www.emas-register.de

Since then, the number of participants has been decreasing while criticism of the scheme has been increasing. Although Germany has still the highest numbers of participants compared to other EU member states, the management system is not widespread within industry and the public sector. As a consequence, two questions remain: first, what are the reasons for participating in EMAS and its effects within organisations, also considering other EMS standards (BMU/UBA 2005) – in the case of this study especially the ones regarding public authorities? Second, how effective are NEPIs in general and EMAS in particular? Some authors question the system in general and want to develop it into a CSR-strategy (Loew/Clausen 2005) while others want to continue with it and therefore make it more attractive to potentially interested organisations (Umweltgutachterausschuss 2005).

3.12 Conclusion of the chapter

The development of EMAS from the first ideas to the first EU regulation, its revision to EMAS II and the present revision towards EMAS III has shown that a wide range of actors have had an influence on the regulation. While the management approach was supported by the UK and other nations, the technical approach was brought in by Germany. In addition, it has become clear that national developments of EMS have also guided the development of the EMAS regulation as well as its use in the member states.

Regarding the implementation of the regulation in national and sub-national structures, this reflects the overall politics and policy structures of both the UK and Germany. While in the UK, the top-down structure was a common practice and therefore went quickly and smoothly, in Germany the federal set of institutions together with the idea of participation of relevant societal groups made it more difficult to set up national structures. Nevertheless, EMAS was and is by far more accepted and used in Germany than in the UK. This is due to a number of factors. A central one is the political and financial support through various programmes, primarily initiated by sub-national actors that make organisations aware of EMAS and help them to implement it. The number of participants in the Länder reflects their activities. For public authorities, a large number of projects and initiatives have been established and Germany has the largest number of public authorities registered under EMAS in Europe. Nevertheless, the EMAS community is still small compared to the number of organisations using ISO 14001 or even to those who do not have a registered EMS at all.

In the UK, the number of participants in the European system has always been low, not only compared to other EMS systems like the BS standard or ISO, but also compared to other countries in the EU. First, this is due to almost no activities for the promotion of the management system, second there is a general trend for organisations use the ISO standard and third there is a considerable lower public and private interest in environmental activities compared to other countries, despite the fact that EMAS as a policy instrument resembles the policy style that is widely used in the UK. In regard to public authorities, only one organisation has actively promoted the implementation of EMAS, focusing on the local level.

The development of the European environmental management standard and its implementation in the two countries give no sufficient answers to the effect the system has on an individual organisation. Therefore, this will be the main focus of this work.

4 Theoretical Concept

4.1 Introduction

There is no genuine theory to analyse the implementation of EMAS within public administration. Therefore, after the definition of the central relevant terms for this chapter, three different theoretical approaches are elaborated. The first is the public administration theory, with the classical Weberian Model of administration on the one hand and the management and resource-orientated idea of operating public administration under New Public Management on the other hand. The second approach is the policy cycle model, which is used to analyse the introduction and implementation of EMAS as a cross-sectoral policy instrument. Third, the elements of change, learning and continuous improvements which EMAS contains are taken into account with an organisational learning approach. Finally, these three elements will form a theoretical concept which, in the end, will work as the background for both the method and the analysis chapters.

4.2 Organisation – definition of a central term

EMAS, in the sense of the policy instrument, is an environmental management tool that helps organisations to analyse and decrease their environmental impacts. It is merely an intra-organisational management system like the quality-focused ISO 9001 or its international EMS rival ISO 14001. Therefore, the definition of the term 'organisation' is central to the argumentation in this work.[46]

Within classical sociological theory, the term 'organisation' comprises three different categories. First, an organisation can be seen as a social process, i.e.

46 From an etymological point of view, the term 'organisation' derives both from the new Latin term 'organisatio' that was created in the 14th century, meaning the creation of an organism and from 'organisare' meaning "creating, developing and ordering of something". Within the French revolution, the word 'organiser' was derived from it, having the meaning of creating a whole thing, developing something, creating something according to a plan. From there, the German and English term 'organisation' was developed and used to describe institutions made by the people that would be important entities of the state, the economy and the society (Grolcha 1969, p. 192, Walter-Busch 1996, p. 24).

as a process of organising something, or rather social processes because usually, there are numerous processes running parallel to each other. In this sense, organisation is an activity rather than an institution. This terminology is often used by economists but they often do not consider the social dimensions and aspects of the activities that are undertaken in the process of organising something. Instead, they focus on the final goal or target of the activity (Schanz 1992, Mayntz 1963, Endruweit 2004). Second, 'organisation' can be understood as a sociological object, working as a social catalyst for processes which changes through an intervening variable. This definition originates from social anthropology and reflects long-term relationships within a society. Thus, society is organised and has a social structure (Endruweit 2004). Third, from a sociological point of view, 'organisation' can be defined as a social subject which is an object of sociological analysis. This social subject or social entity itself can act and develop its own identity. This definition is most common when talking about an organisation because it reflects that an organisation is more than its parts and that within an organisation, there are social processes that are developed only and especially through the organisation's structural framework (Endruweit 2004, Schanz 1992, Mayntz 1963, p. 147). 'Organisation' can also be seen as a specific historic form or a specific relationship of society for the constitution, the establishment and the reproduction of dominant societal developments of the modern world (Türk/Lemke/Bruch 2002, p. 10).

Generally, an organisation inherits the following attributes that create interdependencies amongst each other:

1. Targets: these targets are defined as a 'mission statement' or 'aim-advantage concept', which are the leading ideas for the development of the structure of the organisation. These targets are changing constantly due to influences within or without the organisation or new constellations of power of the members of the organisation. The formal targets of an organisation are usually designed to be long-lasting and laid down in official documents like laws, orders or other formal documents. Next to these formal and visible targets, there are informal secondary targets, one of them being the development and the security of the organisation itself. From these two layers of targets, conflicts can derive, for example if the self-preservation of the organisation is seen as being more important than the formal organisational targets. The most important instrument to achieve the targets is a set of rules that defines tasks, authorisation and activities of the members and units of the organisation (Endruweit 2004).

2. Social Structure: all social interactions within an organisation can be described as the social structure. This structure covers both the formal as well as the informal structure. The formal structure is the planned allocation of tasks and competences in order to give clear rules and positions to the members of the organisation. These formalities constitute functional activities and the hierarchical structure of the organisation. Whereas the functional layer can be seen as a horizontal structure, the hierarchical dimension is a vertical structure. Both structures have an official character and are usually well documented (Zimmermann 2000, p. 261). The informal structure can be characterised as the structure of behaviour. It covers the actual behaviour of the members of the organisation, the relationships among them as well as the group structures. These informal social structures develop due to the needs of the individuals to establish and further develop social contacts. These structures represent the interests, capabilities and motivation of the members of the organisation and are a separate net-like construction that can overlay the formal structures as well as neutralise or amend them. Both structures are in a reciprocal dimension to each other.

3. Surrounding environment: organisations do not work and operate all alone but are embedded in a surrounding environment. They develop relationships of interaction with other institutions; it is like an input-output framework. This environment is, on the one hand, the physical environment, where the organisation takes resources like water, air, energy, etc. for consumption and change. On the other hand, organisations are embedded in their social environment. Consequently, organisations are part of something, i.e. part of administration, the economy, state, or the political system, on local, regional, national or international level (Gukenbiehl 2000, p. 153).

In order to get a positive definition of organisation, one needs to find the characteristics of an organisation. These are as follows:

- the orientation towards specific targets
- the creation of it to explicitly achieve those targets
- a formalised structure
- hierarchy and fixed descriptions of duties of its members
- its lasting for a certain period of time
- the control of it through centres of power
- optimisation of the members of the organisation through exchange of members or workforce
- rational coordination of activity
- exactly definable members and procedure for entrance to organisation

or, put differently, an organisation is a complex system of interaction and cooperation (Endruweit 2004).

Of course, different disciplines look at the term 'organisation' from other angles but within the context of this work, it is seen from a sociological point of view, taking the organisation as a social subject that has durable and formalised structures and that can act within its environment. It can be distinguished from other social subjects. To clearly distinguish it from other terms, one has to bear in mind that there are numerous forms of social systems. There are large ones like society, class, or tribe as well as small systems like family, a group of friends or the colleagues in one's own department, which have, as social systems, no specific single aims or targets. Further, they do not usually have a special form of structure or hierarchy. Therefore, one can say that an organisation is a social system with a very specific structure and with very specific targets. This definition of the term 'organisation' does not take into account the instrumental dimension of the term, which understands the organisation as a tool of the organisational leaders to execute their planned activities, because the aim of this study is to analyse the processes of introduction and implementation of a management system.

To distinguish the term 'organisation' in the sense in which it is used here, one can say that an organisation is not a network, a virtual organisation or a social movement. These are also terms that are used to describe forms of organisation but they are not suitable here because they do not inherit all the structural issues mentioned above. Networks can be defined as "a patterned set of relationships among actors or groups in a social space" (Bacharach/Lawler 1981, p. 205), i.e. they are social systems where their elements have established certain relationships. However, networks are not as goal-orientated, nor do they have a very specific hierarchical structure like organisations do; they are merely informal groups with – compared to an organisation – less formal structure. They are often seen as an alternative layer of coordination which exists above, next to, under or in other already established structures. Sometimes virtual networks are characterised as organisations as well. Therefore, the term has become very popular within the last years; it was brought into the discussion of the organisational theory by Davidow/Malone 2002. Within this form of cooperation, a number of independent organisations work together and thereby bring their main competences together. For the outside customer, they appear as one unit or organisation. Because it is a form of cooperation, it can be seen as a social process that is, as mentioned above, not sufficient enough to be defined as 'organisation'. Furthermore, one can say that an organisation is not a social movement because the latter is defined as

> "collective challenges based on common purposes and social solidarities, in sustained interaction with elites, opponents and authorities" (Tarrow 1998, p. 4).

If one looks at social movements like the workers' movement, the youth culture or other forms of social movements, it becomes obvious that these do not have a very specific structure as mentioned above. They are merely groups of people that are largely un-institutionalised. When such a group becomes, for example, a political party, a church or a union, is it transformed into an organisation. This then is defined as

> "[...] a complex or formal organization which identifies its goals with the preferences of a social movement or a countermovement and attempts to implement those goals" (McCarthy/Zald 1977, p. 1218).

Nevertheless, according to the definition of social movements, these are political organisations which struggle with the existing political order. Consequently, the distinction between social movements and other phenomena like political parties and interest groups is not always sharp (Staggenborg 2008, p. 5).

4.3 Public administration as a special form of organisation

Whereas the description above was to give a general definition of the term 'organisation', public administrations are a special kind of organisation, in contrast to private organisations, like companies, clubs or associations. In the following, these specifics will be outlined in a general way. The term 'administration' implies two aspects, a functional one and an institutional one. The functional dimension of administration describes the office work or the activity of planning, organising and controlling processes. Seen from the institutional point of view, administration can be described as the executive power of a state – next to the legislative and the judicative powers. Within this executive power, one can differentiate further between a 'government' that decides on the political programmes and an 'administration' that executes the will of government (Eichhorn et. al. 2003, p. 1107). The organisations summarised under 'Public administration' are a complex network (Mayntz 1985, p. 1) that operate within the bureaucratic network of organisations of a state.

The responsibility of an administration depends on its function within the setting of administrative organisations and their tasks and on the territorial structure it is responsible for. Within this study, the regional structures of Germany and England are important.[47] At present, Germany is generally divided into three administrative levels: federal administration, state (German: Land) administration and local administration, consisting of district councils, city councils and town councils. In the UK, there is the central government, the administration for the four home nations or constituent countries as well as regional and local administration. Historically, all four nations were divided into counties. These are no longer the sole units for local administration – all four parts of the UK differ in their local administrative structure.

For England, which is the relevant home nation within this study, there are currently nine Government Office Regions. These regions are subdivided into counties and unitary authorities. Whereas the counties represent the traditional division of England and are them-selves subdivided into districts, the unitary authorities combine both administrative functions of county and district. Within large urban areas, this form of single-tier authority is called Metropolitan Council. The capital is divided into boroughs, which are relatively similar to the unitary authorities.[48]

Public administration as a special type of organisation is, in its modern form, i.e. from the 19th century onwards, guided by general principles which have been noted by Max Weber, a rational systems theorist who believed that bureaucracy was the most efficient form of organisation and essential for modern society. Weber saw modern administration in the sense of bureau-

47 Within the EU, the member states are divided into at least three subdivisions for statistical reasons. These levels, called NUTS, are orientated towards the total inhabitants as well as towards negotiations about the size of these levels between the European Commission and the member states. Germany is divided into up to three NUTS levels, NUTS 1 being the sub-national Länder level, NUTS 2 the regional Länder level and NUTS 3 the local level. The large states are divided into three levels, the small ones (Berlin, Bremen, Brandenburg, Hamburg, Saarland) only have NUTS 1 and 3. The UK is divided into up to five NUTS levels, only the first three are of relevance here. NUTS 1 level is a large regional division which is not represented by any administrative division. NUTS 2 is a subdivision of NUTS 1 and again a formal comprehension of several counties for statistical reasons. NUTS 3 is the county level.

48 Since the 1990ies, the division of counties and districts has been changed in some areas. Up to that time, both levels of administration had their tasks. Because this structure was not seen as effective enough, both authority levels have merged into unitary authorities (Becker 2002, p. 95 ff).

cratic administration as an important element of legal rule.[49] First of all, admiistration has to work according to the rules that have been set by the relevant competent body. Second, it is important that there are generally systematic competences of the various authorities and that these competences are set and clear for all that have to deal with these organisations. Third, it is absolutely necessary to have well-qualified civil service staff that are independent from other resources and influences and work according to the organisational hierarchy and structure. Their work is done within offices and has to be documented; documents have to be collected in files to have a record of the administration's activity. The administration's work has to be done without interference of personal interests or with regard to the status of any person. It has to be free from any political influence and concentrate only on the rationale of the administrative work (Weber 1980). These elements of organisation are traditionally considered to be essential for a modern administration.

One can, of course, argue that Weber's definition of public administration is rooted especially in the Prussian administrative system. Although this is correct regarding the changes that public administrations have undergone within the last decades, Weber's definition defines the major principles that have been and still are general guidelines for public administration.

4.4 Differences in political systems of the UK and Germany with consequences for public administration

When comparing public administration in the UK and Germany, one has to keep in mind that there are two different historical developments that determine the major factors of policy within these two countries. In the following, a brief outline is given of the major differences between the political systems

49 Weber distinguished between three different types of authority: bureaucratic, charismatic and traditional. Charismatic authority is based on the unique personal qualities of an exceptional individual, such as a religious prophet or magnetic political leader, rather than any established institutional position, election or office. The authority of traditional political systems such as monarchies is based on long-standing and seldom-questioned, often sacred principles like the hereditary superiority of nobles, religious position or status or other reasons not necessarily related to a person's ability to rule. This form of authority usually has a great stability but also clear limitations because it is not based on the technical qualifications of effectiveness of those in position to make decisions (Weber 1980).

that have significant effects on how public administration in the two countries works.

First of all, one can state that there has been political continuity in Britain whereas in Germany, there has been political rupture. With the development of the parliamentary system of Britain, which began in the 13th century and has continued unprecedented until today, the country has not suffered any major disruption or invasion since 1066. This is contrasted by Germany's history, which is marked by numerous political and territorial ruptures, from the beginnings of national unity, the Kaiserreich, the inter-war years, the Nazi-regime, the Cold War period until Germany's reunification in 1990.

A second major difference between the two countries can be described as formally unlimited versus constitutionally restricted powers of Parliament. Whereas the British Parliament has sovereignty to do „everything that is not naturally impossible“ (Rose 1982, p. 52), called "ultra vires" because in the absence of a written constitution, it developed a tradition that was carried by convention, precedents and practices and therefore has the power to deal with every aspect it wants to (Slapper/Kelly 2001, p. 10), the German constitutional history is marked by a struggle for parliamentary supremacy and the function of the written constitution that was a major factor in Germany's history. This long-lasting struggle is presently reflected by the fact that the fundamental rights in the constitution (German: Grundgesetz), articles 1-20, can never be changed, not even by a majority vote of parliament (Möller 2004).

A third tradition or major difference is the idea of the state itself and the question of what principles the law should follow. Whereas in the continental European mindset, the idea that the state would be a separate legal entity from society was developed, this concept is contrary to the English tradition where the idea of the state is orientated towards individual institutions that, in turn, get their powers from Parliament. This goes in line with the tradition of the Common Law that is based on custom, usage and the decisions of the law courts through case law.

The German legal traditions have their origin in Roman law, in which the state was a separate legal entity based on the constitution. This constitution defines the state's powers and tasks that are executed by the state's 'organs', like parliament, government or other institutions. Nevertheless, certain areas of policy deviate from these traditional ideas of state in both countries. While in Germany, with its strong state tradition, large areas of social services have been delivered not by the state but by NGOs, in Britain, regardless of the 'stateless tradition', the provision of public services was characterised by a predominance of direct public delivery until some years ago (Wollmann 2000, p. 6).

Together with the development of a modern state and the different legal traditions, the position of the state within society was formed. In Britain, the influence of utilitarianism is clearly visible in the fact that the state is seen as being instrumental to the society. In contrast, with the rise of the absolutist terrtorial rulers in Germany, the state was seen as legality in its own right, standing above and beyond partisan interests. This idea became even more prominent with the emerging of the welfare state model in the 1970ies.

As already mentioned, the two countries have completely different traditions of the rule of law, which, in turn, has consequences on the way public administration works. Because Britain has not developed an elaborate system of administrative law compared to the countries on the continent (Slapper/Kelly 2001), there is no separate system of administrative courts and, consequently, no large need for legally trained personnel within the public sector. Government, or should one say governance, be it local or central, tends to prefer the Oxbridge-educated generalists. As a consequence, Britain has not developed a special education programme for people working in the public sector (Norton 2001).[50]

In Germany, the legal system developed into the German constitutional state (German: Rechtsstaat), meaning the elaboration of administrative law, which is designed to regulate the relation between the citizen and the state in all sectors of state administration. Therefore, the administrative law became ever more detailed, corresponding with the need to develop a system of administrative courts. The German 'Rechtsstaat' as it is today is strongly influenced by the developments of the Nazi period and its criminal excesses of the law. Thus, it is seen as the highest constitutional duty to govern according to the law to the highest degree (Benda 1995). Parallel to the administration law, the demand for legally-trained staff increased and has led to a 'monopoly of

50 Thus, Norton (2001) remarks that starting with the reforms under Margaret Thatcher, a greater emphasis on managerial and business skills has been developed. In addition, Norton (2001) sees a less rigid structure within the central government, leading to changed communication and interaction structures within departments.

lawyers' (Juristenmonopol) in public administration until the present time (Schwanke/Elbinger 2006).

Another major difference between Germany and the UK is the administrative structures in both countries, which can be described as unitary or centralist in the UK and decentralist or federalist in Germany.[51] Historically, the 'United Kingdom of Great Britain and Northern Ireland' developed from combining four nations together that differ considerably, for example in the nature of the law, comparing England and Scotland. The constitutional development within this process shows some interesting features: while one can argue that the UK has a unitary government, one has to keep in mind that the Crown came to 'preside over a multi-national kingdom' (Rose 1982, p. 6) and that, as a consequence, there has been no development of an explicit regional level of government.[52]

In contrast, the constitutional and institutional development in Germany led to a federal structure – with the exception of the Third Reich. After the Second World War, the German states (Länder) were (re-)established followed by the rebuilding of a federal structure with a two-chamber parliamentary structure which led to a 'unitary federal state' (Hesse 1962), in which power is divided between the federal government, the states and local government in regard to the different policies.

The major differences that have an effect on public administration in the two countries have been summarised in the following table.

---

51 Regarding the Federal Republic of Germany and, after re-unification, Germany as a whole.

52 The Government Offices of the Regions set up in 1994 are not seen as a regional government structure because they primarily are part of the central government in London. Nevertheless, there is a clear development of the devolution process within the UK. While the Labour government gave more powers to the Scottish with the re-establishment of the Scottish Parliament in 1999 and also to a lesser extend to the Welsh with the establishment of the Welsh Assembly in 1999 it is more cautious to take on the process in England, let alone in Northern Ireland. As a response to the development of devolution on the continent and as a reaction to EU policies, the central government established Regional Development Authorities (RDA) in 1999. These are responsible for the economic, and social development within their region, they are to raise the average prosperity to that of the rest of the European Union (Copus 2001). Thus, devolution is progressing; it is by far not as detailed and explicit as in federal states.

Aspect of divergence	UK	Germany
Development of political culture	Political continuity	Political rupture
Power of Parliament	Formally unlimited, no constitution	Restricted by constitution
Vision of the state	State-less, based on Common Law	State tradition, based on Roman Law
Rule of Law	No administration law, no need for specially trained staff	Detailed administration law, need for specially trained staff (Juristenmonopol)
Administrative structures	Unitary/Centralist	Federalist/Decentralised

Table 4.1: Differences in regard to public administration in the UK and Germany. Source: Own.

All these elements determine the work of public administration in the two countries and have consequences regarding the policy instrument implementation. The political culture is determining the political activities while Parliament – in this case, regardless of the supranational EU powers because the concentration is on the national level here – provides the legal framework for the use of policy instruments. The visions of state as well as the rule of law again determine the work of executive and legislative power while the administrative structures give a framework for the operation of public authorities.

4.5 Building a theoretical framework

After the outline of the major differences that affect the comparative analysis of the British and German local government policy, politics and polity, in the following, the theoretical concept of this work is elaborated. Due to the fact that there is no single comprehensive theory for the purpose, it is developed out of the following three elements: public administration theory, the policy cycle and organisational learning.

4.5.1 Theory of public administration

There is not a single theory of public administration, but numerous clusters of theories, like the ones of political control of bureaucracy, the public institutional theory, the theories of public management, the network theory, etc.

Within this study, the concentration lies on the classical theory of bureaucracy, as developed by Max Weber, on the one hand and the theories of New Public Management on the other hand. Both represent the ideas of public administration (theory) of their time and in their oppositeness, both are the predominant guidelines for public administration in the two countries looked at in detail.

4.5.1.1 Classical theory of public administration

Beginning with Max Weber and Henry Fayol, the classical theory of public administration has developed a set of principles. First of all, it is important to separate politics and administration. Politics should concentrate on the goals while the operative tasks should be given to a politically non-active administration. Second, the organisation has a hierarchical structure which is part of the command and control mechanism (Weber 1980, Handel 2003, Ritz 2005).

Weber's model of a rational administration is orientated towards the concept of legal rule, which is based on the following criteria:

- the division of labour with a highly differentiated overall structure
- the fixed and standardised structure guided by a set of documented rules
- the hierarchical structure with a chain of command and obedience
- the division of identity between function and person
- the specially-trained staff
- the lifelong career and employment for civil servants
- the specific work ethos for the public administration.

According to Weber, this ideal type of bureaucracy is superior to two other forms of power he identifies, the traditional rule or the charismatic rule. If bureaucracy follows these principles, it becomes rational, efficient and accountable. Weber argued on the basis of ideas of industrialisation in the 19th century. In a way, he follows the Prussian model of administration and exaggerates it (Reichard 1995, p. 58, Nitschke 1998, p. 168).

Every form of rule is part of a process of socialisation. Ideally, this process leads to a rational use of power. In this case, public administration is the best example of a rational regime (Nitschke 1998, p. 165). The main factors of this regime are the following:

- a system of order or regulation, made by man through rationality
- a system of force and

- a form of organisation (Weber 1980, p. 466).

When the system of order and regulation, together with the system of force, come together within the appropriate form of organisation, then it is possible to talk about bureaucracy in a Weberian sense. The use of power does not only work with force but is a matter of liability and law.

Despite the enormous influence that Weber's model of bureaucracy had in the past and still has today, one should not be mistaken in thinking that Weber saw his model as related to practice. It has to be seen as an analytical category, a kind of ideal type to justify rational power in the light of his time. Thus his idea of rational power was the starting point for the development of the modern age (Siedentopf 1976, p. 14, Nitschke 1998, p. 166). Furthermore, Weber's model is relevant for the general and structural questions of public administration, concentrating on the ideal type of organisation rather than on its daily work (Nitschke 1998).

Whereas in the classical theory of administration, the normative description of how administrations have to work and which activities can be undertaken to increase their efficiency are predominant, the neo-classical theory of administration concentrated more on the realistic description which was done in order to explain why political programmes developed as they did; sometimes relating to the legislatively-intended effects, but producing unexpected and even problematic side-effects.

4.5.1.2 Neo-classical theory of public administration

The discussion about a theory of Public Administration developed into four fields from the 1960ies onwards, each with a different focus (Jann 1986, Bogumil/Jann 2005), thus creating the neoclassical string of Public Administration theory:[53]

- Moorstein-Marx's book "Public Administration" (1965), which is a description of the structures and functions of administration based on empirical evidence,
- Thieme's book on "Verwaltungslehre" (1966, 4th ed. 1984), which comes more from the juridical viewpoint of Public Administration and

53 The neo-classical theory of public administration was influenced by a number of theories and ideas, such as the early human-relations-movement, the theory of decision by Herbert Simon as well as the systems theory.

provides practical information on the structures of the German administration, in particular,
- Maier's book on "Die Ältere deutsche Staats- und Verwaltungslehre (Politikwissenschaft)" (1966, 2nd ed. 1980), which referred back to the older German „Staatswissenschaft" and, finally,
- Ellwein with his "Einführung in die Regierungs- und Verwaltungslehre" (1966), providing an explicitly empirical view with the focus on political functions and the increase of power of administration.[54]

Based on the classical theory, the neo-classical view has a wider focus because it takes the social situation of the administration's staff into account in order to increase the organisation's productivity. Therefore, it tries to build up the necessary structures of communication as well as giving incentives for staff to improve motivation. This theoretical string was biased by economical theories, mainly the welfare economics with its three main functions: allocation, distribution and stabilisation.[55] Additionally, the systems theory was taken up by the neo-classical string of public administration theory and widened its perspective. Through the systems theory, it was made clear that a larger number of sciences are needed to analyse and change public management because it is a rather complicated form of organisation. Thus, through the definition of interfaces, it is possible to reduce the complexity of these systems. Under the systems theory, public administration can be further analysed when dividing large systems into sub-systems. By doing so, it is easier to analyse their relationships with other sub-systems and with the surrounding environment. These findings lead to a concentration on structures and functions of the actors within the organisations.

4.5.1.3 Model of Public Administration under New Public Management

The term "New Public Management" (NPM) is a synonym for numerous reform strategies that are imposed on Public Administration. NPM's focus is on the micro-economisation of public administration. It does not offer a catalogue of measures that have to be undertaken but wants to introduce a

54 Next to the German "Verwaltungstheorie" is "Verwaltungslehre", which one could translate as Public Administration Science. This is more orientated towards the education of Public Administration's staff, thus producing textbooks and publications and giving practical advice.

55 These are still very dominant in German politics, less in British politics where welfare economics has been pushed back by M. Thatcher and the following government leaders.

new vision and mission in public administrations: efficiency and effectiveness of all tasks that are in focus. This implies a greater responsibility for all members of staff as well as an orientation towards free market principles and competition. One can structure the NPM concepts along the following two perspectives:

- A macro dimension, which tries to reduce the activities of the state (central, regional and local level, if applicable) to their core activities. This includes the relationship between administrations as well as the relationship with their addressees or customers. This is called the inter-organisational perspective.
- A micro dimension, which has a focus on the activities within the organisation itself. New concepts of organising, human relations, internal management and communication techniques are used to change the organisation from a traditional Weberian bureaucratic style towards a management style that resembles private business organisations. This dimension is called the intra-organisational perspective.

The development of NPM is a result of numerous changes in politics, society and economy in many countries around the world. Despite the fact that the situation is different in every country, there was a significant correspondence in regard to public sector reform (Borons/Grüning 1998, p. 12), especially in the Anglo-American world.[56] The global economic crisis that began in the 1980ies lead to a decrease in public finance which, in turn, developed into a question of what should and could be financed by the public sector. The political support for an expansive welfare state declined and at the same time the change within many societies supported a change of values from a community that was dominated by duty and acceptance towards a lifestyle of individual self-fulfilment. This post-materialistic view changes all areas of life. Parallel to these developments, neo-conservative and ordo-liberal positions came up, which were popular in the 1980ies with the development of

56 Despite the fact that the NPM concept has been taken up by numerous countries, there is a general debate about the transferability of management concepts from one cultural system to another. Among comparative management researchers, this question is debated between the 'universalists', and the 'culturalists'. The 'unversalits' are of the opinion that management methods are independent of the cultural and country-specific context in which they are applied (Osborne/Gaebler 1994). The so-called 'culturalists' take the opposite view, saying that cultural norms and patterns of behaviour have an importance when applying management methods. According to this theory, different cultural contexts require different management methods (Kickert/Jorgensen 1995).

Thatcherism and Reagonomics. Generally, these political positions were predestined for a reduction of public administration activities rather than an overall reform of the public service sector.

The theoretical bases for these developments were the public choice theory regarding the general discussion of public administration as well as the new institutional economics for the intra-organisational questions of public administration reform (Damkowski/Precht 1995, Ritz 2003, Bogumil/Jann 2005). The public choice theory tries to explain the decision processes in politics with the help of ideas from micro economics. According the theory, individuals behave rationally and benefit-orientated with a focus on their own individual advantage. This behaviour leads to structural deficits in the political system because the "political market" works imperfectly so that there are large (financial and structural) losses within the political and social welfare systems. As a consequence, individual preferences of citizens are disregarded because there are not enough possibilities to choose from in the public sector (Buchanan/Tullock 1962, Buchanan 2003). Nevertheless, there is a group of powerful actors that can use this system for its own benefit. Therefore, the system has to be changed towards a stronger perspective on user interests, which include more possibilities to participate and a greater choice of services as well as more competition in the public sector.

A second string of theories are those of institutional economics, mainly theories of transaction costs (Williamson 1987). The last is often used to argue in favour of privatising public services because a central argument of the theories of transaction costs is that private companies would have more interest in working efficiently with the money they get for a service compared to public administration. The principal-agent-theory states that relationships between individuals in organisations are seen as principal-agent relationships that can work with contracts. The theory of transaction costs puts the focus on the question of providing a service from within the organisation or from outside. The guideline for decisions is the price of goods and services. The general question is whether to deliver the goods and services from within the organisation or to buy them from a contractor. According to the principal-agent-theory, an organisation should concentrate on the core competences that it can produce at market-competitive prices and give all other goods and services to contractors (Holzinger 2007). A third line of approach is the use of management methods like the "management by objectives" idea and the use of controlling instruments.

Based on this set of theories, core themes of the NPM concepts are the following:

- limitation of activities within the public sector to those tasks that cannot be done by private companies at the same quality at less cost,
- use of management methods of the private sector with special focus on responsibility of leading managers, giving them enough freedom to do their job,
- work according to management by objectives and clear indicators,
- change of structures, dissolving monolithically organised structures, building of new decentralised flexible structures of (partially) independent units that are run by contract management,
- use of competition to improve service quality and financial discipline,
- customer orientation, citizens are treated like customers, not like petitioners (Damkowski/Precht 1995, Borins/Grüning 1998, Lane 2000, Schröter/Wollmann 2001, Ritz 2003, Thom/Ritz 2004).

Although the NPM concept has been used and discussed for almost two decades, there is a debate about the transferability of management concepts from the private to the public sector. Beginning in the late 1990ies, the ideas of NPM came under criticism, thus leading to a re-orientation of public administration towards a concept of the "Neo-Weberian State" (Pollitt/ Bouckaert 2000). Additionally, there is severe criticism about the implementtation processes, the instruments used and the unintentional side effects of NPM because the concept tends to focus on instruments rather than on structural changes, although the criticism of NPM depends on the depth and length of its implementation as well as on country specifics (Pollitt/Bouckaert 2000).[57]

4.5.1.4 Comparing bureaucracy and management model

If one compares both the classical bureaucracy and the modern management concept, it becomes clear that the concepts ideally lead to different modes of operation of public administration among a large number of issues.

57 This criticism is more detailed in Germany than in the UK, despite the fact that NPM or its German variant NSM have not been as widely implemented compared to the UK (Jann et. al. 2004, Bogumil et. al. 2005).

Bureaucracy	Management
Aim: efficiency	Aim: efficiency
Separation of administration and politics	Separation of administration and politics
Guiding principle: hierarchy	Guiding principle: market
Input orientation	Outcome orientation
Legality	Legality and efficiency
Stability	Adaptation
Conditional programming (if-then)	Final programming (what-how)
Transparency and documentation of procedures	Transparency, efficiency and documentation of procedures, new methods of business economics
In case of problems: Legal process	In case of problems: termination of business and/or legal process
Officer has to work in an independent and impersonal way	Officer has a self interest
Hierarchical, unitary structure, monocratic leadership and accountability	Flat hierarchy, more room for manoeuvre and less control, strengthening of output-responsibility for leading personnel, new reporting systems for politics and administration
Specially trained staff for a very detailed task, principle of speciality and differentiation of labour in order to have an efficient administration, experts working together	Expansion of staff education, especially in business economics, focus more on processes, less on functions, teams of different backgrounds working together
Principle of the unity of division of labour, clear structures, each member of staff has one superior officer	Support of target and task orientation, management by objectives
Delegation of tasks in a top-down principle	Management by objectives, controlling and reporting instruments
Reduction of party influence in connection with staff selection	Responsibility for staff lies with the top management of the administration

Table 4.2: Comparing bureaucracy model and management model of public administration. Source: Own, based on Reinermann 2000, p. 20 and Ritz 2003, p. 118.

The bureaucracy model is primarily the Weberian ideal of an administration and is the result of industrial society's development, which required an administration based on the principles of accountability, equity and legality. Within this model, public administration is part of the state executive, providing stable, steady and reliable services for people and private organisations. In contrast, the management model derives from the private sector and is influenced by a wide range of managerial ideas mainly focused on the outcomes of operation. It is merely a financially-driven service provision and delivery with a strong tendency to be cost-cautious.

Both models are not free from problems. The bureaucracy model of administration is more a model or an ideal type that would be a model for development of the modern world (Nitschke 1998, p. 166). As with all models, there is the question of the difference between ideal type and reality. The NPM-model is a problem-orientated toolbox for the internal reform of organisations with a strong focus on outcome instruments (Pollitt/Bouckaert 2000, Bogumil/Jann 2005), with instruments or solutions often not fitting to the needs of administrations or being rejected by staff as well as the general public. Nevertheless, both models have some common elements. First, their aim is to modernise the state. Second, they reflect the political, economical and societal developments of their time. While the Weberian Model was a reaction to the feudal system, the NPM toolbox was developed as a reaction to the upcoming problems of the Keynesian welfare state.

4.5.2 The policy cycle

One of the most prominent models for analysing policy processes[58] is the policy cycle, which was developed from the 1950ies onwards. Its main idea is that policy-making is a process in which problems are to be discussed, solving strategies are to be developed, alternative solutions are considered and, finally, the solutions are to be laid down as obligatory for all participant actors (Scharpf 1973, p. 15). The reason for developing a new model to explain policy processes was the criticism that the political science community had mainly been concentrating on the input of the political system, i.e. on demand and support of the political system that was reflected by studies about elections, parties, elites, parliaments, etc. but not on the question of the output

58 Policy processes are a central issue of the policy analysis, which, according to a classical definition by Dye, wants to find out "what governments do, why they do it, and what difference it makes" (Dye 1976, p. 1).

of the political system, i.e. on laws, programmes, budgets, administrative or political activities etc. With the discussion about the policy cycle, research began to focus more on the decisions and actions, rather than on demands and supports. The discussion about policy lead to a new definition of the three dimensions of political science, the policy, polity and policy triangle.[59]

4.5.2.1 Development of the policy cycle concept

The concept of the policy cycle defines political processes as a follow-up of phases of problem processes. Originally invented by Lasswell (1956), who wanted to create a new policy science with an interdisciplinary, multi-method approach (Lasswell 1968, p. 182) that would provide information and knowledge about and for the decision process, he suggested a distinction of the process of policy making into the following phases:

1. intelligence – the collection and processing of knowledge
2. promotion – promotion of policy alternatives
3. prescription – the commitment to the selected decision
4. invocation – the enforcement of a political decision or policy
5. application – to put the policy into action within the administration
6. termination – the termination of the policy and finally
7. appraisal – evaluation against the planned goals and perspectives.

---

59 This was especially new for the German political science perspective. However, the new distinction helped to focus on the policy process, i.e. on the dimension that tackled the process of policy-making. According to Scharpf, policy analysis can be used for the following:

"Analysis of this kind can contribute to the ability to decide within a democracy; it is possible to show which problems have been repressed, which targets have been neglected and which alternatives of action have already been excluded during the process of preparing a decision by the administration before any responsible politician was occupied with a proposal to decide. Analysis of this kind does not need to end at the level of the formal decision of the legislative body or a secretary of state but can be extended towards the implementation phase. Thus, they can depict how many questions have not been decided upon through the formal political decision and how the smaller or larger frame of activity is being filled or changed by administration", (Scharpf 1973, p. 16, translation by author).

It was pushed forward by the fact that at the beginning of the 1980ies, whenthe deficits of many political programmes became obvious, the concepts of political planning had failed due to problems of implementation. Therefore, the analysis of the policy process and especially the focus on policy implementation seemed to be a logical step.

Lasswell's model seemed relatively successful, especially in the USA of the 1960ies, when policy research projects began which were developed to a larger extent through several reform ideas, financially backed by the Ford Foundation. These projects let to the development of new study programmes focussing on policy implementation (Jann/Wegrich 2003, Schneider/Janning 2006). The first model of policy making was developed within the 1970ies (Brewer 1974, Anderson 1975, Jenkins 1978, May/Wildavsky 1978). Mayntz (1977) introduced it to the German discussion. Though different in terminology and focus, all models are going in the same direction.

Rational planning and decision	Lasswell 1956	Brewer 1974	Anderson 1975	Mayntz 1977	Jenkins 1978	May/ Wildavski 1978	Standard Model used here
Problem formulation and targets		Initiation	Agenda Setting	Problem articulation, target definition	Initiation	Agenda Setting	Problem perception
Information management	Intelligence	Estimation			Information	Issue Analysis	Agenda setting
Generating of alternatives	Promotion						Policy formulation
Comparison and assessment	Prescription		Formulation		Consideration		
Decision	Invocation	Selection	Adoption	Programme development	Decision	Service Delivery Systems	Decision making
Realisation	Application	Implementation	Implementation	Implementation	Implementation	Implementation	Implementation
Control	Appraisal	Evaluation	Evaluation	Evaluation	Evaluation	Utilization of Evaluation	Evaluation
Termination	Termination	Termination		Termination	Termination		Termination

Table 4.3: Types of the policy cycle model. Source: after Jann/Wegrich 2003, p. 77 and own additions.

Within this study, the model of the last column is used. It represents the classical model of the policy cycle (Windhoff-Héritier 1987, Prittwitz 1994, Howlett/Ramesh 2003), incorporating all relevant phases, thus combining the phases 'generating of alternatives' as well as 'comparison and assessment' in one phase that comprehends the whole process of policy formulation. This is done for two reasons: first, because it is very difficult in reality to get data to distinguish between the two phases; second, because with the phases proposed in the last column, it fits perfectly to the EMAS model described in chapter 3, resembling the PDCA-cycle as well as the idea of continuous improvement of EMAS.

In theory, the policy cycle model is a sequential decision process (see figure 4.1). It remains to be seen within the analysis of the EMAS processes whether these sequences will be visible and dividable in a clear form.

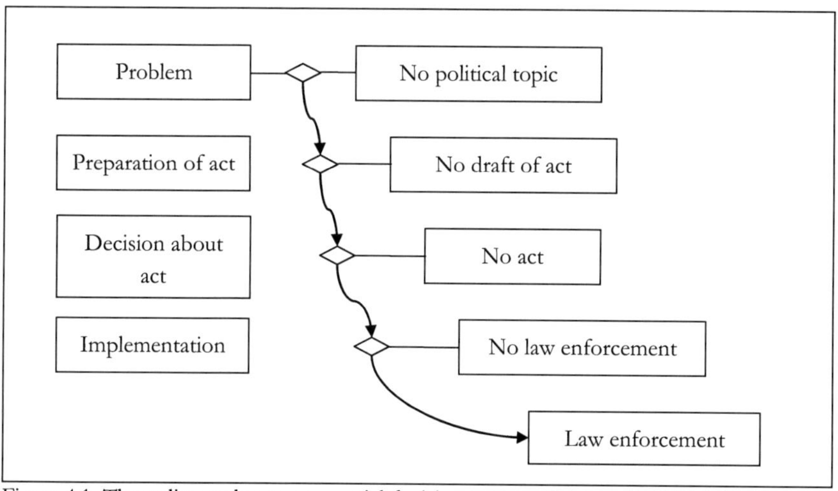

Figure 4.1: The policy cycle as a sequential decision process. Source: Schneider/Janning 2006, p. 116.

4.5.2.2 The phases of the policy cycle[60]

Problem perception: political problems are collectively recognised problems. According to Sjöblom (1986), a problem occurs when there is a discrepancy between something that there is and something that is aimed for. Only

60 The policy cycle and its phases are described here because they contain terms that are used within the data analysis of chapter 6.

through the recognition of a problem by the relevant political actors it is possible to get them to focus on it. However, one has to keep in mind that this perception process is very selective. Therefore, the media or other public attention is often an important key of starting the perception of a problem at all. The recognition of a problem together with the agenda setting is the start of the policy cycle (Windhoff-Héritier 1991, p. 63, Jann/Wegrich 2003, p. 83).

Agenda setting: Regarding the question of how agenda setting works, there are several theoretical approaches. From a system theory point of view, there is an implicit logic of development of this phase: The socio-economic development of society and its crisis will automatically lead to a point where the political system has to react to the upcoming problems (Zapf 1969). This approach regards the conditions for the agenda setting process as secondary and postulates a logic of development that automatically leads to problems being regarded as worth being on the agenda. A second approach follows the line of capacity building, i.e. that the agenda setting can only start when there are capacities available for problem solving (Prittwitz 1990). A third perspective is more focused on the current political situation. Based on the garbage-can-theory (Cohen/March/Olsen 1972), Kingdon (1995) assumes that there are three independent "streams" of activity. If these streams meet, there will be a policy window or 'window of opportunity' to get the problem on the agenda. One of these streams is the 'problem stream', which includes information about the problems that are in need for a political solution. A second stream is that of activities by relevant authorities that have the knowledge and experts to solve the problem. These experts have – according to the garbage-can-theory – often already analysed the problem and proposed solutions which have not been put into practice so far. Both streams need a third one, the 'political stream', i.e. the attention, interest and power of the politicians to set it on the agenda. This theory is the opposite of the system theory because there is no implicit logic but an active involvement of actors needed to get a problem on the agenda.

All these theories stress the importance of the three main factors, media, public and politics, which have a continuously rising importance and interdependence to get a problem on the political agenda. However, it is not only a question of how something gets on the agenda but also, vice versa, a question of how good the political agenda is based on the public discourse. Another aspect of agenda setting is the question of the direction of the initiative for this agenda setting. Ideally, the impulse can come from a bottom-up-perspective through society or from a top-down-perspective through (sub-national, national or supra-national public) administration. According to Jänicke (1999), agenda setting, especially in environmental policy, is either innovative or adoptive. Innovation in these terms means environmental innovation; that is the market introduction of a product that has a disburdening effect on the environment. This can happen through end-of-pipe technology, a reduction or substitution of material needed for the product or an increase in the eco-efficiency of the product. These innovations are a subset of eco-innovations (Jänicke 2008). Thus, there are four variants of agenda setting that are:

- policy innovation by public authorities
- policy adoption by public authorities
- policy innovation by society
- policy adoption by society

Innovation processes originate as a result of generation and application of knowledge. Generally, one can distinguish between four characteristics of the innovation process:

- the opportunity to innovate in terms of the acquisition of knowledge
- the distribution of resources available to invest in innovation
- the distribution of incentives to innovate and, finally,
- the capabilities of the relevant organisations to form innovations strategies and manage the process of innovation (Dosi 1997, Andersen/ Metcalfe/Tether 2000, Freier 2005).

Policy adoption, in contrast, is a process of up taking of already existing ideas and knowledge for own purposes.

Policy formulation: In this phase of the policy cycle, political programmes are discussed and formulated. The term "programmes" is used as a general term for targets, instruments, strategies, responsibilities, finances, etc. This phase is very much dominated by the struggle for the solution to the problem. Within this process, it is important to equalise or to try to balance all actors' interests. Policy formulation is a central point in the political process. The results and activities of this phase are the policy outputs. On the one hand, these can be primarily symbolic activities like the decision to appoint a new working group for a specific topic, which can have no great financial or political implications. On the other hand, the activities can imply concrete and precise problem solving with a time frame, financial resources and a system to evaluate the progress of the action. Sometimes, this phase is analysed together with the policy decision phase (e.g. in Jann/Wegrich 2003, p. 85) because both phases interact – especially if one has a number of actors that have to come to a decision on a programme. However, for analytical reasons, it seems more practical to distinguish between both, especially within public administration, where according to the traditional model of bureaucracy, staff members who formulate a policy as the executives do not decide on it.

Central for policy formulation is the frame in which the problem and its solution are situated. Often, certain policy options are not possible due to the limitation of resources, i.e. financial, personnel or other restraints (Windhoff-Héritier 1987, p. 84-95). Another important aspect is the individual influence of each actor within the relevant arena, which does not only include the actor's competence and responsibility but also the distribution of power among the actors. To a certain and growing extent, policy consultants can play a central role in the process of policy formulation (Lathrop 2003).[61]

Decision making: A decision is always a compulsory choice between several alternatives which cannot be realised altogether at one time. These decisions are taken by the relevant persons or authorities that have the power to do so. Within policy analysis, the focus at this phase is not so much on the characteristics of the persons that decide or on actual situation of decision, but more on the empirical evidence of a decision that takes the realities, complexities and dynamics of decision making into account. This automatically leads not only to the formal deciders, but to the actors that make the preliminary decision because it has become clear that persons in certain policy networks who make these preliminary decisions tend to direct the policy process on a specific track. Furthermore, the closeness or plurality of policy networks has been put in the spotlight (Windhoff-Héritier 1987) and it has become clear that these actors can have a most important influence, ranging from the dominance of the policy field through interest groups to a strong influence of administration towards a pluralistic system of activity (Howlett/Ramesh 2003, p. 125).

Policy implementation: After the decision has been made by the relevant actors, this phase is where the decision is turned into activities; the policy programme that was decided upon begins. First of all, it has to be made clear who is responsible for the programme and who works with it. Second, the

61 This is especially the case with EMAS or with other EU policy norms. Though it is more often the case in the UK to employ consultants within public administration for special tasks, German administration has begun to use these experts widely as well, especially regarding reform programmes of the administration.

allocation of resources is essential for the following activities. Third, it is important to know how to decide in special cases.

The policy outcomes and the policy impacts that are generated within this phase are a result of both the decision and the implementation phase. Both outcome and impact have to be seen together because one implies the other; this is called a policy-impact analysis (Schneider/Janning 2006, p. 61).

There are two ways of looking at implementation processes: one is prescriptive implementation research, which assumes a homogeneous will and clear preferences leading to definitive activities. In turn, the realisation of implementation problems leads to an improved planning of future programmes in order to solve future problems. Obviously, this approach is very difficult to conduct because often the goals that were set in the policy formulating phase cannot be reached due to a large number of different circumstances which are not clearly visible or foreseeable, like the open and hidden interests of the actors or political influences of other actors. Further, it is often not easy to find the borderline between external conditions and institutional or processorientated conditions that influence the implementation (Jann/Wegrich 2007 p. 89-92).

The other way of looking at implementation processes is descriptive implementation research, orientated towards a reflection on the implementation process using empirical and analytical methods without any focus of future improvements. The strings of activity are followed, the actors, their activities as well as the interdependences are analysed and thus, a total dismissal of goal-orientation is not possible. To summarise, one can say that there is no total, but a gradual distinction between the two forms of implementation research: one is more goal-orientated, the other one more process-orientated. For the analysis of the EMAS processes within this work, the second perspective that concentrates on the processes is dominant.

Another important element of the implementation phase is the actors (Windhoff-Héritier 1987, p. 92). Generally, there are three groups of actors: first, the ones that formulate a programme, the so-called 'programme conceptionalists', second, the ones that have to carry out the programme – these are separated into the two groups of 'implementations managers' and the people who really do the work ('street bureaucrats' in Lipsky's terms, Lipsky

1980, p. 391). Third, there are the addressees or 'target groups' of the programme.[62]

A critical problem of the implementation phase is the coordination of the activities of the different actors. There are several possibilities to do this. First, one can try to define common goals that have to be achieved; this is called 'mutual adaptation' (Mc Laughlin 1978). Second, it is possible to achieve a solution through negotiation, which leads to a situation where common targets for all actors are defined. Third, it is possible to have a power scenario where a possible conflict is solved by the one actor forcing others to follow his preferences. This leads to an extremely uncooperative behaviour of the other actors while one actor can only fulfil his goals at the cost of the others.
When a programme is being implemented, this process can be seen from a top-down-perspective that focuses on the implementation of the programme or from a bottom-up perspective that questions the role of the hierarchical, central regulation in order to get more flexibility with the current problems of implementation. Research has shown that it is important to keep both perspectives in mind, but that in practice they are often interrelated. At the beginning of implementation research, the top-down-perspective was predominant; much analysis concentrated on the preciseness of target achievement. This was based on a very hierarchical view of political processes (Mayntz 1987). Later, the perspective was changed to a more interactive policy style (Wollmann 1983).

A programme always has implications on other programmes. Therefore, it is important to look at the relation of one programme towards others. One can see positive and negative intertwinement of programmes, which can be important for each programme's development (Windhoff-Héritier 1987, p. 104). Regarding environmental policies, these intertwinements are important because, usually, cross-sectional or multi-dimensional problems can be solved with those programmes.

62 A special focus of implementation research has been the fact that organisations themselves have a tendency to concentrate on their own benefit rather than on that of the addressees of the programme. Organisations have vital interests in securing their influence on power and resources in order not to be abolished. This does not only count for whole organisations, but also for individuals within these organisations. Thus, a great factor of policy implementation is not only the outcome but also the personal benefit of it. These factors will also be part of the analysis.

Policy Evaluation: The starting point of the policy cycle model was the question to what extent the set goals and impacts have had an impact. Furthermore, the effects of the impact are important. Therefore, in this phase, the policy cycle model has close links with evaluation research. There are several methods to evaluate the outcomes:

- ex post or summative evaluation looks at the effects and successes of a programme after its termination
- ex ante evaluation involves a pre-assessment of the impacts and side effects of forthcoming activity. One of the most prominent instruments for this type of evaluation is a cost-benefit-analysis[63]
- an evaluative pre-assessment is used to find out if a programme or activity is suitable for evaluation
- formative or on-going evaluation accompanies the ongoing programme in order to find problems as soon as possible. The problem solutions are used in a back loop to improve the ongoing programme
- descriptive-analytical observation or monitoring without causal interpretation or analysis that is mostly done with standardised indicators.
- Furthermore, there are evaluations of the effects and the efficiency of a programme. The first looks at the grade of target achievement, the second focuses more on the relation between inputs and results (Wollmann 2003).

Consequently, the policy evaluation phase is an important but at the same time difficult and critical process because, at this point, it becomes obvious whether a policy programme has been a failure or a success. The results of the evaluation can be used in the political arena by several actors for their political and strategic purposes.

With the development of NPM, policy evaluation in the form of financial control under an outcome perspective has become an important tool for assessing policy instruments and programmes. Nevertheless, evaluation of policy had already, in some way, taken place before that, whether in the political arena with the assessment of the output by political actors and the public or within the administration itself through financial control institutions, courts or government papers. Now that programme evaluation is explicit and politically necessary, it is even more visible and dominant in regard to the

---

63 Within the NPM concept, controlling and performance measurement play an important role (Blanke et. al. 2001, Wollmann 2003, Kuhlmann/Bogumil/Wollmann 2004).

continuation or termination of a policy programme. The policy evaluation can lead to changes in the political process or the current policy programme, which can be described as political learning (Bandelow 2003, p. 289-331).

Policy Termination: The idea of termination is that a policy programme has solved a problem and that it can therefore be terminated (Bandelow 2003). Besides this, a programme can be terminated for other reasons, as the results of the evaluation phase have indicated, although this is not usually the case. More often, financial constraints, new policy windows, a change of government or changes of central actors lead to a policy termination. Further, a policy termination always has to be seen in the light of the relationship of power of the relevant actors. Therefore, the following aspects, which are signs of a weak programme, can lead to a policy termination:

- the cross-sectoral perspective of the programme
- the duration of a policy problem
- the social stigmatisation of a problem or
- the novelty or the absent tradition of a programme (Windhoff-Héritier 1987, p. 108).

Often, policy programmes are not terminated at once but are extended, changed into new programmes or phased out slowly, especially when one can expect large support from important actors for a programme.

4.5.2.3 Criticism to the policy cycle model

With the rising number of studies that have used the policy cycle as a model for analysing policy programmes together with more fundamental criticism (Sabatier 1993, Héritier 1993, Prätorius 1997, Sabatier 1999, Schneider/ Janning 2006), one can sum up the major points as follows:

- From a descriptive point of view, the segmentation of the policy process into logical and sequential steps cannot be seen as such. Empirical research has shown that the steps are weaved together and cannot be identified separately. Sometimes some of them are left out or changed in their order.
- The policy process cannot be described as a cycle model because it is more a sequence-model and, therefore, the policy process only continues if a positive decision towards the next phase is made. This view is especially prominent when regarding actor-orientated models. Nevertheless, this position is not regarded as sufficient within this study because it does not consider the fact that policy instruments are usually

changed or terminated after a certain time. Thus, this model is missing the aspect of continuity that many policy processes have.

- From a conceptual point of view, the policy cycle is not a theory because there a no definitive variables that explain the transition from one phase to the other. To analyse the different phases of the policy model, one has to develop singular theoretical frames that are usually independent from each other and fail to identify the model as a whole.
- Additionally, it was often underestimated that a lot of activities in politics have more of a symbolic character rather than being put into practice. Programmes and activities are set up to secure power and influence. The problem solving can, therefore, be of secondary nature.

As a consequence, the policy cycle model has been seen as oversimplified and unrealistic by its critics. Therefore, other models to analyse policy processes have been favoured in the discussion. Most prominent amongst them are the "Advocacy Coalition Framework" by Sabatier, the "Institutional Rational Choice", "Policy Diffusion" and the "Funnel of Causality" (Sabatier 1999). Nevertheless, even the critics of the policy cycle model do not deny its importance as a heuristic model, i.e. as a conceptional tool to find out specifics of the policy process (Schneider/Janning 2006, p. 65, Jann/Wegrich 2007).

4.5.2.4 Relevance for the study

Despite the criticism of the policy cycle model, it is highly relevant for this study because it helps to get a differentiated understanding of the internal dynamics, the peculiarities and the causalities of the specific and complex processes of policy making (Mayntz 1983). It contributes to the knowledge of policy processes and helps to logically structure those processes which are analysed within this study. Furthermore, it is a tool to reflect policy processes. In addition, the policy cycle discussion, together with research about NPM and other factors, has lead to the current debate about the concept of governance (Benz et. al. 2007). Despite its critics, it is still a useful tool to segment policy processes and to make them visible.

Even if the model has been developed to structure and identify policy processes within a policy field or regarding a policy programme with numerous actors and institutions with an inter-organisational perspective, for this study, it is used with an intra-organisational perspective. The policy cycle model is the underlying structure for the analysis of the policy processes that leads to the introduction and implementation of EMAS, which itself is taken here as a cross-sectoral example of an intra-organisational management tool, introduced in a number of public administrations throughout Europe as a reaction to European legislation. Additionally, EMAS fulfils the criteria for a new management instrument introduced within the NPM-process of modernising public administration. Finally, the policy field of environmental policy has been under a lot of pressure during the last decades due to the arising necessity to develop it further and has also been faced with numerous problems resulting from this cross-sectoral perspective. These three dimensions, the European one that affects local policy dimensions, the reform process of public administration under NPM-influence and the cross-sectoral view of environmental policy, will be analysed with the help of the policy cycle by generating data regarding the policy processes that took place while introducing and implementing the policy instrument as well as by evaluating these processes. Methodologically, the policy cycle is the framework for the questionnaire-guided interviews that gave the empirical database for this study because it structures the policy process that took place within the organisation (see chapter five for details).

4.5.3 Organisational learning

4.5.3.1 Introduction

While the last two parts of this chapter concentrated on public administration and the policy cycle, the following part will draw attention to the theory of organisational learning by elaborating a model of organisational learning which will contribute to the continuous improvement process that is inherent to EMAS. Public administration reform is about change. Under the NPM-model it is a change towards a better use of resources and delivering of services. Change, therefore, has something to do with a changed behaviour of people working within or for that organisation in order to achieve the goals of better resource allocation and service delivery. This changed behaviour can very broadly be defined as a form of learning.

To start with, the term 'learning' or even 'organisational learning' has quite different connotations. Learning in the context of organisational science is seen as new views and 'new knowledge' (Agyris/Schön 1978, Hedberg 1981), 'new structures of the organisation' (Chandler 1962, Wolff 1982), 'new methods and systems' (Jelinek 1979, Shrivastava 1983), 'new standard procedures' (Cyert/March 1963, Miller/Friesen 1980), 'new strategies'

(Mintzberg/McHugh 1985), 'new cultures' (Smirchich 1983, Schein 1985) or a combination of these aspects (e,g. Bartunek 1984, Pautzke 1989). Additionally, there is an open debate about the used terminology within the field of organisational learning which ranges from 'learning' to 'adaption' and from 'change' to 'forget'. Another major issue is the question of the factors that influence organisational learning. While the discussion is often about structures, strategies and organisational culture, the question of their combined impact remains insufficiently investigated.

A further aspect which has to be considered is the fact that organisations consist of individuals. These individuals interact within an organisation. Consequently, the question arises as to whether these individuals or the organisation itself learn:

> "There is something paradoxical here. Organizations are not merely collections of individuals, yet there is no organization without such collections. Similarly, organizational learning is not merely individual learning, yet organizations learn only through the experience and actions of individuals" (Argyris/Schön 1978, p. 9).[64]

The task of the following is to clarify this paradox. First two major theories which influence individual learning theories are discussed. After that, as the second part, theories of organisational learning are examined in detail.

4.5.3.2 Individual learning

Learning processes of an individual can be distinguished by the behaviouristic and cognitive theories of learning. Behaviourism originated from behavioural psychology; learning is described as a change of visible behaviour, based on logical positivism, which states that there is an objectivity and comparability of observation (Popper/Eccles 1989, p. 91). Learning in this sense is a changed behaviour as a reaction to a stimulus. The stimulus can be the same for a number of learning processes. Learning according to this theory works strongly with positive and negative stimuli.

Neo-behaviourist theories go a step further and use interceding variables or hypothetical constructs that cannot be exactly observed to explain learning procedures. The hypothetical constructs especially are seen as arranging

---

64 Thus, the idea that organisations, rather than its individual members can learn and that this process is more than the learning process of its members is often questioned (Wiswede 1991).

variables that help one to look closely at human thinking and its problem-solving capacity (Bower/Hilgard 1983).

The cognitive theories of individual learning focus on the mental work that an individual conducts when it learns. According to constructivism, cognitive processes of learning lead the individual to knowledge of their environment and to the ability to control this knowledge (Kroeber-Riel 1990, p. 218). Central for the learning process in this sense is the acquisition of new knowledge and the ability to put it into contexts in order to solve problems (Bower/Hilgard 1983, p. 37, Kroeber-Riel 1990, p. 344). Whereas among behaviourists, where the learning process took place within a "black box" that only made visible the reactions after someone had learned, the constructivists try to look for the reasons for behaviour, for the fact that through learning reactions and routines, cognitive structures are developed. These structures help the individual to cope with new information. Learning in this sense can be
a) an increase in knowledge that is added
b) a development of the cognitive structures so that they are more detailed
or c) the re-organisation of structures due to new information that leads to new knowledge. Therefore, the individual is in a constant interaction with their environment based on experience, expectation and conviction and the already-acquired knowledge structures of their environment (Bandura 1979, p. 22). Additionally important for learning processes is the learning based on a model, which leads humans to the ability not only to learn according to their own experience but also to adopt ideas and experiences from others. In this sense, learning does not have to be clearly visible in a changed reaction, but it is essential for changed behaviour:

"Change resulting from learning need not be visibly behavioural. Learning may result in new and significant insights and awareness that dictate no behavioural change. In this sense, the crucial element in learning is that the organism be consciously aware of the differences and alternatives and have consciously chosen one of these alternatives. The choice may be not to reconstruct behaviour, but, rather, to change one's cognitive maps of understandings" (Friedlander 1983, p. 194).[65]

Although both theories of individual learning are not totally incompatible (Gagné 1970), there are a number of reasons to focus on the cognitive approach:

---

65 From a systems theory point of view, learning is described as "the term that one cannot observe how information leads to far reaching consequences because they lead to partial structural changes in a system without breaking the self-identification of the system." (Luhmann 1984, p. 159, translation by the author)

1. A large number of the experiments that lead to the behaviouristic theory of learning have been done through animal testing. The forms of learning done by these animals are at least a very simple phenomenon of learning which cannot be used to explain the complexity of human learning processes.
2. Behaviourism presupposes that the associated or conditioned reaction is already included in the repertoire of behaviour of the organism. Therefore, the aim of the theory is not the explanation of changed patterns of behaviour but the explanation of the probability of occurrence of already known patterns of behaviour in a situation of stimulation. Thus, the behaviourist theory cannot explain the development of new behaviour, which is utmost important.
3. Behaviourists think that changes of behaviour are caused by external enforcement. Consequently, the situation of learning is reduced to the solution of a known task under the lead of a guiding person. Regarding organisational reality, this does not reflect the learning environment at large.
4. The necessity to reflect and to develop anticipative thinking is essential for the development of problem solving capacities in the field of environmental policy. Therefore, behaviouristic methods of learning do not seem appropriate (Bandura 1979, p. 31, Pautzke 1989).

The following table summarises the differences between behaviourist and cognitive learning theory.

	Behaviourism	**Cognitivism**
Subject of learning	Individual	Individual
Object of learning	Stimulus-response	Problem-solving
Result of learning	Change of behaviour	Change of cognition
Relation of individual towards environment	Individual reacts to stimulus from environment	Creative response to environment
Theoretical background	Positivism	Constructivism

Table 4.4: Differences between behaviourism and cognitivism. Source: Own.

For the task of analysing learning processes within organisations, using a policy instrument that implies constant improvement and a reaction to new situations, the cognitive approach seems the right one to explain organisational learning because it fits both to the models of the EMAS continuous cycle of improvement as well as to the policy cycle.

4.5.3.3 *Organisational learning*

Organisational learning can be seen from two different angles: one is that it is seen as learning of individuals within an organisation, the other is that one tries to identify processes within an organisation that explain learning within it beyond individual learning processes (Pautzke 1989). Some authors even think that learning is primarily done by the organisational elite (de Geus 1988, p. 70), whereby individual learning theories are transferred to organisational learning.[66] Nevertheless, it has become clear that it is neither possible nor useful to have learning processes only at the top management level (Malik/Probst 1981, p. 138). If one takes into account that the knowledge of an organisation is spread over all levels of it, it becomes clear that learning at the top level only is not appropriate. Therefore, it is necessary to promote learning processes on all levels of an organisation to achieve the best result possible.

Generally, there are three types of organisational learning that go beyond an individual view of learning:

- Learning of organisations through the formalisation of individual learning experiences which leads to an independence of the organisation in regard to its individual members who underwent these learning processes.
- Learning as a change of the knowledge that is shared by all members of organisation and
- Organisational learning as the use, change and further development of the organisational knowledge base.

These three types will be discussed in the following. The first to develop the theory that organisational learning was possible through the formalisation of individual learning experiences was Taylor.[67] Through observation and

66 Usually, a three step model is used to explain this process: 1. definition of types of practices towards learning of the top management, 2. transferability and transfer methods of these practices to the organisation and 3. deduction of characteristic organisational reactions to changes of environment (Heimel-Wagner 2003, p. 54).

67 Interestingly, Taylor worked to increase efficiency, like Weber, at about the same time, though; he had industry in mind and not public administration. Furthermore, he did not want to analyse the consequences of this process of rationalisation on society, like

documentation, he turned individual knowledge of workers into knowledge for the organisation. Consequently, it was possible to replicate the knowledge of the individual worker, so that increasingly less-qualified personnel could be hired. The formalisation of individual knowledge can also lead to a change of standard procedures (Cyert/March 1963), a change in methods and systems (Jelinek 1979), structures (Chandler 1962) or strategies (Mintzberg/McHugh 1985). Overall, this model of organisational learning inherits a clear cut between individual and organisational learning. However, through its formalisation process, this model does not take into account the changes of knowledge, especially if it is not documented. Further, this rather "industrial-age" approach is very much concentrated on division of labour and does not consider motivational aspects of work. As these aspects have a growing importance when describing organisational learning today, it becomes clear that this form of organisational learning is not sufficient any more, at least not for this study.

A second approach for a model of organisational learning, abstracted from the individual, is the view that organisational learning is a change of the knowledge that is shared by all members of organisation. Jelinek (1979, p. 16) describes this as the "shared frames of reference", Duncan and Weiss (Duncan/Weiss 1979, p. 86) use the term "communicable, consensual and integrated knowledge" whereas Argyris and Schön (1978, p. 17) use the metaphor of "organizational maps" to describe this phenomenon. This approach has a focus on the organisational culture, which is primarily the complete knowledge of all members of organisation. Here, all members of an organisation are involved; therefore, an organisation must make learning on all levels of it a high priority. Critics of this approach claim that the assumption of a homogeneous culture within an organisation and, consequently, a common knowledge does not reflect the reality of organisations in general (Schreyögg/Papenheim 1988). The pluralisms of form and style of life that are predominant in post-industrial societies influence the culture of organisations. Therefore, the individuals within organisations have a large amount of specific knowledge that is only accessible through the individual member.
A third approach to organisational learning was developed by Kirsch (1990). He defines organisational learning as a process of usage, change and further development of the organisational knowledge. Based on the knowledge that is potentially available through its members, organisational learning takes place if the members place their knowledge at the organisation's disposal or when

Weber did (Reichard 1995, p. 58). Nevertheless, the aim of both was the same, to develop models of organisation in order to work more rationally.

organisational knowledge is developed further. These developments can only take place if there is an environment for learning, i.e. hindrances for learning have been removed. This rather wide approach takes into account three different levels of organisation: the individual, a group within the organisation or the whole organisation as such.

In the end, organisational learning is based on the individuals; their learning processes[68] are necessary to change an institution. Without these processes, there can be no change within organisations. Therefore, the content of organisational learning is a change of the organisational framework which makes learning processes possible for the individual. On the one hand, there are knowledge bases that are not accessible to the organisation; on the other hand, there is information that is saved within the organisation through its capabilities (Argyris/Schön 1978). As a consequence, there is a qualitative and a quantitative difference for organisational learning compared to individual learning because organisational learning is characterised by a collective rationality. This means that the organisation sets up a collective framework, defines norms of behaviour and, by doing so, demands a process of adjustment of the individual towards the organisation.

To conclude, within the context of EMAS and its implementation in public administrations, the following definition of learning seems to be appropriate:

> "Organisational learning is a process of increasing and changing the organisational basis of values and knowledge, the improvement of the competence to decide and to solve problems as well as a change of the common frame of reference of and by members of an organisation." (Probst/Büchel 1994, p. 17, translation by author)

4.5.3.4 Organisational learning – the concept of Argyris and Schön

Argyris and Schön (1978) developed an approach that takes into account both perspectives of the individual as well as of the organisation.[69] They postulate that human activity or action is based upon the knowledge of the individual and inherits basic norms, strategies and expectations about the consequences

---

68 Individual learning processes are, for example, described with the spiral of learning (Kolb 1984), the PDCA cycle by Deming (1986). The EMAS audit cycle is based on the PDCA-concept by Deming, thus linking between individual and organisational learning.

69 A large number of theories of learning are orientated towards the approach developed by Argyris and Schön (Wahren 1996, p. 43). This cognitive approach of organisational learning is the theoretical model for organisational learning within this work.

of specific behaviour under a fixed set of rules. Within an organisation, one can find two sets of rules which they call 'theories'. On the one hand, there are 'espoused theories' i.e. the set of rules that contains the official way of action and activity. This set of rules proclaims the formal principles of action in an organisation.[70] Their aim is to guide the members of organisation in a certain direction. Often these rules are set through the organisation's vision, mission or other strategic papers or given to the members of the organisation through organisational laws like in the military. On the other hand, one can find 'theories-in-use', i.e. a set of rules that contains the unofficial, informal rules of the organisation that are used within the daily routine.[71] These rules are the dominant ones for the members of organisation as they include the collectively shared expectations, the so-called 'collective knowledge', the norms for the daily routines and details of rules for success and failure. These theories-in-use construct an inter-subjective reality that most members of the organisation share and follow. Every single member of the organisation has a personal theory-in-use; it is transformed into a collective theory-in-use through communication and the positive and negative sanctioning of the individual's behaviour.

The central process of organisational learning is the communication about and the agreement upon theories-in-use by the members of an organisation. These agreements and, in turn, also the theories will be and are in fact changed through the behaviour of the each member of the organisation. The espoused theories, on the contrary, are usually too static to be changed in the short run; they last longer and have a strong formal status. The theories-of-use are developed by individuals, they are represented through a mind map or, as Argyris/Schön (1978) call it, "through private images and public maps". These 'images' are cognitive structures of the individual concerning the theories-in-use while 'maps' are external references that are available to all members of an organisation, like plans, work descriptions, guidelines, diagrams, etc. Together, private images and public maps are the basis for organisational learning:

> "Organizational theory-in-use, continually constructed through individual inquiry, is encoded in private images and in public maps. These are the media of organizational learning" (Argyris/Schön 1978, p. 16).

70 This is similar to the formal social structures mentioned in chapter two when defining the term 'organisation'.

71 This term is similar to the informal social structures mentioned above.

Thus, learning is the cognitive process of knowledge acquisition, knowledge addition and transformation. An organisation learns through its members when they realise that their action, which is guided through the theories-in-use, does not have the effect that they intended. The learning process comes to an end when new knowledge is saved in the organisational collective mind in the form of maps and images.

The learning process itself is seen as an activity in a loop, i.e. as a constant process of change and adaptation. There are three forms of loop-learning; these gradually show an increase in the learning capability which are 'single-loop-learning', 'double-loop-learning' and 'deuteron-learning'.[72]

Single loop learning puts an emphasis on the detection and correction of errors within governing variables:

"[...] members of the organization respond to changes in the internal and external environments of the organisations by detecting errors which they then correct so as to maintain the central features of organisational theories in use." (Argyris/Schön 1978, p. 18)

This learning process is linked to an incremental change within an organisation. New information is added to existing information and the interaction of the members of the organisation with their internal and external environment leads to a constant change of the perceived reality. Learning within this loop is a form of adaptation or correction of their current activity. Existing norms or theory-in-use of the organisation are not questioned at this level; the main task here is the improvement within a given framework.

Double loop learning involves interrogating the governing variables themselves and often leads to larger and more radical changes of an organisation. The basis for this type of learning is the realisation that the changes applied at the single-loop level do not lead to the intended effect.

"We will give the name 'double-loop learning' to those sorts of organisational inquiry which resolve incompatible organizational norms by setting new priorities and weightings of norms or by restructuring themselves together with associated norms and assumptions." (Argyris/Schön 1978, p. 24)

72 See Probst/Büchel 1994, p. 178 for a comprehensive list of forms of organisational learning. It becomes clear that the majority of models follow a three-phase learning system that is similar to that of Argyris/Schön.

Double-loop learning leads to substantial changes that are based on an analysis of the problem. These changes can lead to conflicts within the organisation because the problem-solving process needs more than (simple) adaptation. To solve problems at this level of learning, it is important that the views and theories of the members of the organisation do actually change. Consequently, this involves the re-arrangement of the theories-in-use of the organisation. Therefore, Argyris/Schön use the term 'change learning' for this form of learning. Due to the fact that change learning often leads to personal insecurity, passive and defensive routines will partially be developed by members of the organisation. Although members of the organisation realise the need for change, they often cannot cope with it or allow it due to the effects on their own situation within the organisation. Therefore, processes of change are usually a difficult task.

A third step of learning is the so-called 'deuteron-learning', where the members of organisation learn to reflect on previous contexts and inquire into past situations to as how they came to be.

"They discover what they did that facilitated or prohibited learning, they invent new strategies for learning, they produce these strategies and they evaluate and generalize what they have produced." (Argyris/Schön 1978, p. 34)

The discovery of patterns that have led to changes in similar situations leads to a general reflection about processes and activities. Within this stage of learning, complete re-structuring of patterns of behaviour, strategies and targets of the organisation are undertaken. It is a transformation process that will lead to fundamental changes within the organisation. As well as double-loop learning, these changes can cause insecurity and defensive strategies among the members of the organisation. These insecurities have to be taken into account when thinking about these fundamental changes. This form of overall structural change is usually a top-down process.

One can summarise the three levels of organisational learning in the following figure:

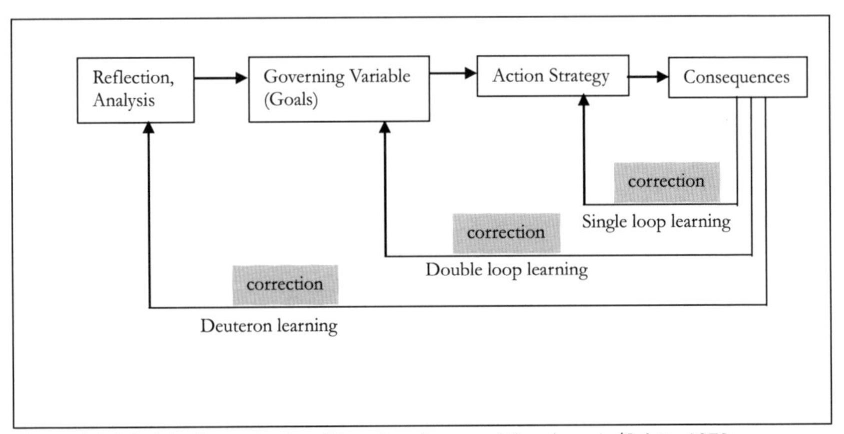

Figure 4.2: Model of organisational learning. Source: After Argyris/Schön 1978.

4.5.3.5 Relevance for the study

With the learning-model developed by Argyris/Schön, it is possible to explain phenomena of change, as described in the definition of an organisation and in the policy cycle model. It corresponds with the idea of the organisation as a social structure and, additionally, reflects the idea of change that is inherited in the policy cycle model. In addition, there is an official line of information, viewpoint and self-image of an organisation that it wants to be transported to the general public. This is especially true for public authorities. Furthermore, there are daily routines and activities inside the organisation which are accessible only for its members. These activities have a very large impact on the organisation's activities in total.

Organisational learning according to the model of Argyris/Schön is a tool that tries to look behind these intra-organisational, informal processes. The implementation of a management system such as EMAS is an interference with the internal formal and non-formal structures of the organisation because large parts of the EMS reflect on the actual behaviour of the organisation's individual members. Additionally, the establishment of such a system affects the formal routines of an organisation because these are questioned. Further, EMAS is a tool that is formal on the one side but informal and, through the PDCA-cycle, always reminding one of the adaptation, change and re-questioning of one's own and the organisation's activities. Therefore, the model of organisational learning – together with the policy-cycle – is the suitable theoretical basis to gain detailed information about the intra-organisational processes that took place while introducing and implementing EMAS.

4.6 Combined theoretical model to analyse the introduction and implementation of EMAS in local public authorities

After the three main theoretical concepts of this work have been outlined in detail, the following part is to create an overall, combined model to analyse the introduction and implementation of EMAS in public administration. This will be the theoretical basis for the analysis of the structurally-guided expert interviews with a number of members of each analysed organisation.

Based on the comparison of public administration of the UK and Germany in chapter two as well as the theory of public administration elaborated in this chapter, generally, two different models of administration are recognised when comparing administrative systems of both countries. One is the traditional Weberian model; the other is the managerially-dominated New Public Management idea with numerous instruments. Both models have implications on how administrations have worked in the past and are working today. Comparing local administration's operations in both countries, the Weberian model is (still) predominant in Germany whereas the NPM model is predominant in the UK, although many of the German administrations have worked with instruments of NPM during the last years. At the same time, UK public administrations have experienced an enormous change beginning in the Thatcher era. Thus, some general principles of the Weberian model are still preserved. While in the UK, the NPM model has been implemented both on national as well as on local level, in Germany the Weberian Model dominates the scene despite numerous projects on national, regional and local level. Public administration reform in Germany is very much dominated by a bottom-up approach due to the federal structure whereas in the UK it has been implemented in a top-down process, enforced by national government policy.

Nevertheless, there are variations to the general rule which will be visible when analysing the implementation of the policy instrument EMAS. The overall theoretical model is a guideline for the individual operation of the organisation and gives a normative framework for the operation of public organisations in general. Clearly, the policy processes that lead to the introduction and implementation of EMAS are connected with the model of administration an organisation is following. These interconnections will be part of the analysis. It will bring to light to what extent the NPM or the Weberian model of administration contributes to the policy process linked to EMAS.

Based on these fundamental distinctions, the policy cycle model is used to analyse the introductory process as well as the implementation of EMAS. This model of policy analysis is sufficient not only for the analysis of policy developments between actors and organisations but also for the analysis of policy processes within organisations because here one finds the same processes as compared to a policy field. The advantage of using the policy cycle within an organisation in the case of this study is the limited number of actors, the clear focus of the organisation and the fact that only a single policy instrument is looked at in detail. In a way, this study is a micro-policy study. Despite the criticism that the policy cycle would lead to an oversimplifying view because it gives the impression that it is only necessary to develop policy programmes and to keep them alive, its structure and idea are ideal for this analysis as they do not only help to separate the steps towards the implementation of the management system, but also resemble the obligatory steps towards a successful validation of the policy instrument, which analysed within this study. With the separation of phases – as in the policy cycle model – it is possible to identify and analyse what otherwise seems to be an unidentifiable and complex process.

In order to look closely at the processes of change that are incremental to the EMAS system, the model of organisational learning is used. The learning processes that take place are looked at in detail on the basis of the espoused theories and the theories-in-use. The EMAS cycle is based on the idea of continuous improvement, which implies processes of organisational learning. In the course of the analysis, it will be possible to see to what extent these learning processes are undertaken, what hindrances are to be found while trying to improve the organisation and to what extent organisations of the public sector are ready and willing to change. One task of the analysis will be to try to categorise these learning processes. The model of organisational learning poses the question of how something is learned in the administration, or to be more precise, what is learned in connection with the introduction or implementation of EMAS. How does an organisation (and its members) learn to introduce and implement a management system? And what kind of benefit will they get out of it? Or, to be more precise: what do these organisations do, why do they do it and what difference does it make? All three theoretical approaches are combined to a theoretical model which provides the basis for the analysis. Figure 4.3 gives an overview of the combined theoretical model used within this work.

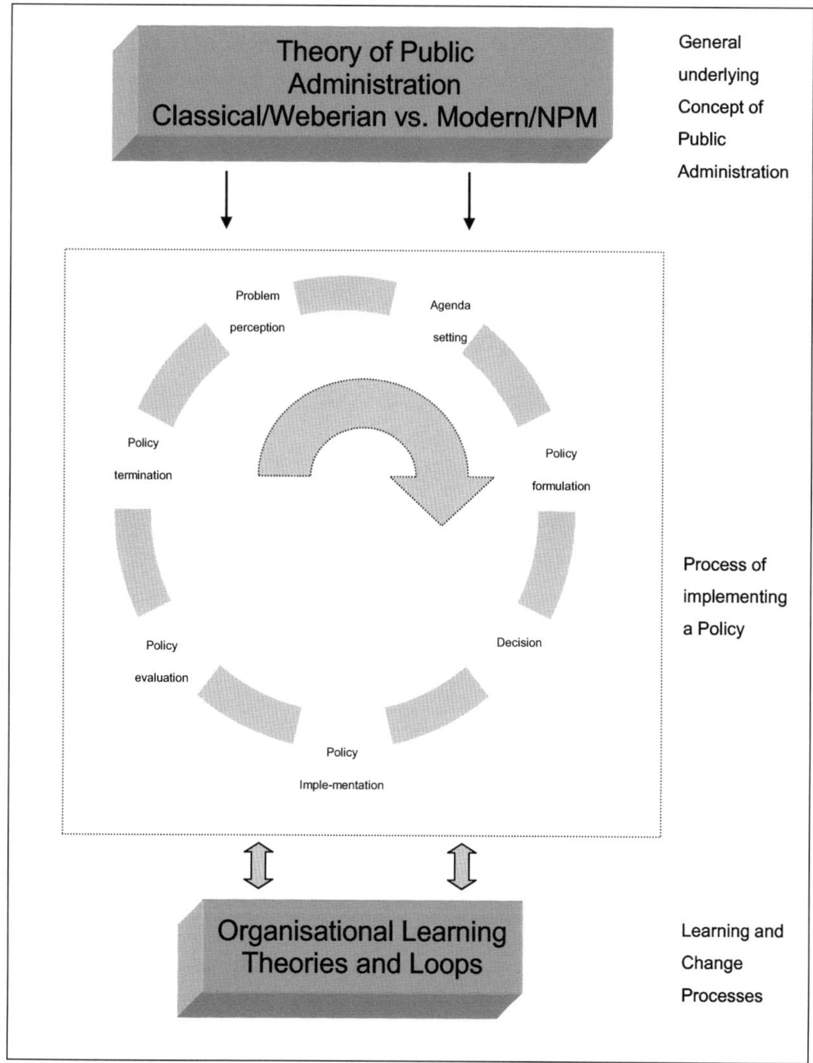

Figure 4.3: Theoretical model to analyse the introduction and implementation of EMAS. Source: Own.

The framework of the model is the organisation, in this case the special type of local public administration in Germany and the UK, while the basis for the analysis is the policy cycle, influenced by the main model of public administration that the organisations follow. The outcomes of the policy evaluation phase will give the basis for the organisational learning analysis. The whole model will lead to a detailed, complex analysis of the introduction and implementation processes of EMAS as well as of the effects that these processes have on the organisation.

5 Case Study Methodology

5.1 Introduction

After a detailed outline of the theoretical model, this chapter is about the methods used, their fields of application and the cases that are examined within this work. The first part of this chapter deals with a discussion of the methods in question and is followed by a description of the sample. The second part deals with the conduction of the interviews as well as the chosen interviewees. Within the third part, the methods of data collection are described. Finally, within the fourth part, a description of the data analysis methods is given. Figure 5.1 gives an overview of the methodical steps of this chapter.

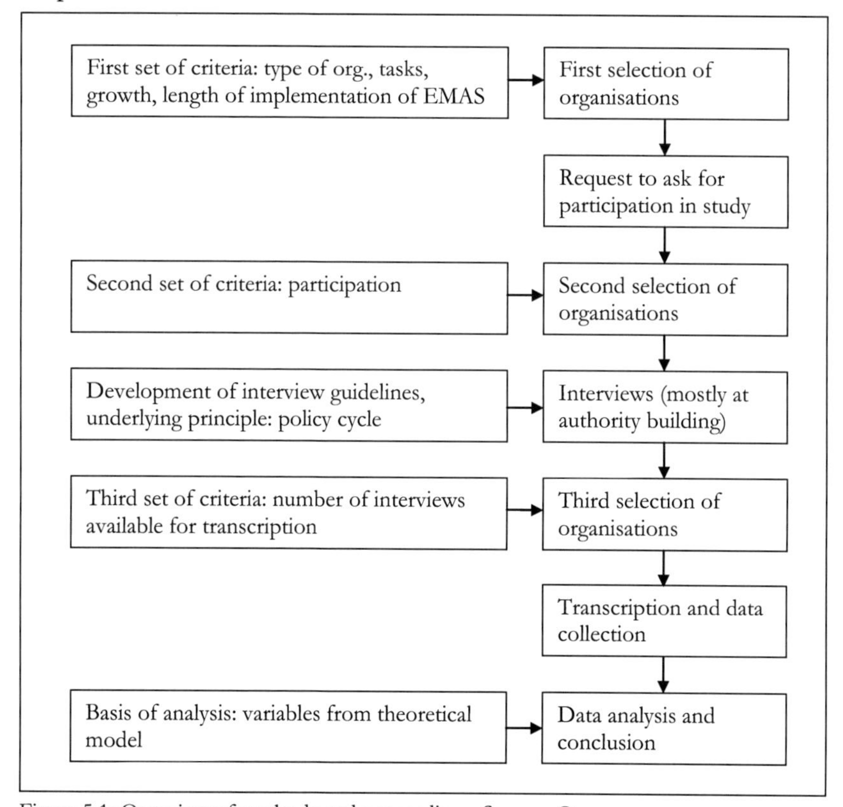

Figure 5.1: Overview of methods and proceedings. Source: Own.

5.2 Choosing the method

This study tries to identify the potentials and pitfalls of the introduction and implementation of EMAS as a policy instrument in local public authorities in the UK and Germany and, therefore, it is mainly an evaluative research design. Evaluation research is a qualitative inquiry with four main aims. First, it checks the effects, efficiency and the achievement of targets of political, social, economical or ecological programmes, instruments and activities. Second, its results give information on how to decide and plan in the future for those who are responsible for the evaluated programme, instrument or policy. Third, evaluation aims to support and encourage learning processes that are seen as relevant for society and politics. It further helps to document these processes. Finally, in regard to an exploratory social research approach, evaluation aims to add new findings to existing knowledge and, thus, enables a deeper understanding of the object or area of research (Kardorff 2003, p. 239). Patton gives a comprehensive and rather broad definition of evaluation:

> "I use the term evaluation quite broadly to include any effort to increase the human effectiveness through systematic data-based inquiry. Human beings are engaged in all kinds of efforts to make the world a better place. These efforts include assessing needs, formulating policies, passing laws, delivering programs, managing people and resources, providing therapy, developing communities, changing organizational culture, educating students, intervening in conflicts and solving problems. In these and other efforts to make the world a better place, the question of whether the people involved are accomplishing what they want to accomplish arises. When one examines and judges accomplishments and effectiveness, one is engaged in evaluation. When this examination of effectiveness is conducted systematically and empirically through careful data collection and thoughtful analysis, one is engaged in evaluation research." (Patton, 1990, p. 11)

Traditionally, evaluation research is the assessment of programmes and organisations. While the first was intended to primarily address social issues and their effectiveness (the so-called programme evaluation), the second focuses on performance and productivity within an organisation (organisational evaluation). A third form is "realistic evaluation". Questioning the above described distinction between programme and organisational evaluation, this form of evaluation uses a model examining phenomena in relation to their context and outcomes. While allowing a rather pluralistic approach to evaluation research in regard to qualitative or quantitative methods, this approach requires a theory-driven basis that can be tested against the data collected (Pawson/Tilley 1997, Kelley 2004, p. 524). This understanding of evaluation research is the one followed within this study.

5.2.1 The interviews

To gather the data needed, problem-centred expert interviews with a number of representatives of the relevant organisations seems the right method to use because it offers the possibility to question the construction or interpretation of situations and to ask about motives of action. Thus, it is possible to get more detailed information about daily routines. Overall, the qualitative approach is favoured over the quantitative because it is not only the frequency of information occurrence that is important but much more the content of this information. Furthermore, the aim of this work is to check the theoretical framework against the empirical material that is filtered through the analysis of the interviews (Mayring 2003, Gläser/Laudel 2004, p. 198). Using the method of extraction as described below, the used variables derive directly from the theoretical model. Thus, within the qualitative analysis, coherence between the subject of the analysis and the theoretical model is given.

The interviews are carried out using a semi-standardised interview guideline (see appendix 1) based on the theoretical model developed in chapter four. Each set of questions has a definite function: the first refers to personal information about the interviewee, i.e. their position and their tasks within the authority, the service or department they are working in and the length of time the interviewee was/has been a member of the authority (question set 1). The second set tackles the introduction and implementation of EMAS within the organisation/unit of the interviewee according to the policy cycle model (question sets 2 to 7). The third set concentrates on the future of EMAS, both within the organisation and beyond, for example in other public authorities and in industry. Additionally, interviewees are asked some questions covering the political future of EMAS (questions set 8).

Although the interviewer used these guidelines, the interviews were conducted as open and semi-standardised interviews (Gläser/Laudel 2004, p. 39), which are characterised by concentrating on a given topic, thus being at the same time relatively free, using the guideline as an overall structure in order to try to get all the relevant information. The interviewees were free to talk about their experiences with and expectations of EMAS without having to strictly observe specific answering categories. Data generated through this method of problem-centred interviews has proved to be

> "[...] honest, more reflected, more precise and open than with the use of a questionnaire or a non-open survey technique" (Mayring 2002, p. 69), translation by author.

This technique also gives one the chance to adapt to the current situation, e.g. a group interview, local characteristics or other special situational cases.

5.2.2 Data collection

The method used here to collect data is the content analysis, which categorises the material in a way that the content is still available but is more structured (Mayring 2003, p. 472). This method is selected because the following three others do not seem appropriate. First, the method of free interpretation, where a researcher interprets the material after reading, is not really a method of analysis because there are no rules of procedure. Thus, this analysis cannot be reproduced. Second, sequential methods like the objective hermeneutic method (Oevermann et. al. 1979, Oevermann 2002) or the narrative analysis (Schütze 1983) try to analyse qualitative data according to the temporal and thematic connection of texts. These methods are very complicated and not very common, though elements of these techniques are used by other methods of analysis. Third, the coding of texts which derived out of the Grounded Theory is another method often used for interview analysis (Strauss/Glaser 1967). Coding is a method to mark text passages that contain relevant information with a shortcut. The codes have either been based on a theoretical model or, alternatively, are developed through the reading process. The result of this analysis process is a system of markings that is spread all over the text. These markings are the basis for further analysis, i.e. the comparison of all parts of the text that contain the same codes or the search for a specific topic in the text. These steps of analysis aim to try and to answer the research question. Coding can be and is often done today with the help of special computer programmes (Kelle 1995, Kelle 2003). This method is not favoured either because often the work is only done with the codes, not with the original texts. In addition, coded structures are often not flexible enough to adapt to new knowledge during the analysis (Gläser/Laudel 2003). Thus, the comprising content analysis as suggested by Laudel/Gläser (2003) is used to analyse the transcripts (see 5.5 and following).

5.3 The sample

5.3.1 The way to the sample

The sample organisations were selected in a two-step process. First, all UK and German public administrations listed as NACE 75.1 were selected from the relevant databases.[73] Information about the organisations was collected and put together in a database to get an overview of the organisations that would potentially be within the sample. The relevant data about the organisations was available through environmental statements, which were either available on the Internet or asked for by post, the Internet pages of the organisations and other publications.

It became clear that in Germany as well as in the UK, a wide variety of organisations were listed under NACE 75.1. In Germany, there are national, regional and local authorities, administrations and organisations ranging from the German Federal Environmental Agency, Berlin, with several thousand employees and an obviously strong environmental background to relatively small local public administrations like the Gemeindeverwaltung Nehren (Town Council Nehren) in the state of Baden-Wuerttemberg with only six employees. Though there are a great variety of organisations in Germany that take part in EMAS, it is obvious that the majority of participants that take part in EMAS are small and medium-sized towns as well as regional administrations (see chapter three for the geographical specifics and other specifics of EMAS within local public authorities in Germany).

The situation is different in the UK. Here, the number of participants is relatively small. On November 1, 2004, the number of NACE 75.1 organisations validated according to EMAS was only sixteen. If we take a closer look at these, it becomes obvious that compared to the German organisations, there is a trend towards larger organisations, i.e. they have a larger number of employees than those in Germany, especially the large unitary city councils like Leeds City Council and Kirklees City Council. However, there are also a number of comparable organisations in the UK. Geographically, there is no focus of organisations having EMAS in one area (see chapter three for the geographical and other specifics of EMAS within local public administrations in the UK).

In a second step, the organisations were selected according a set of criteria. These were the following:

- Tasks of organisation: comparable, work on roughly the same fields of activity[74]

---

73 For the German results it was www.emas-register.de (10 October 2004), the British results come from www.emas.org.uk (01 November 2004).

74 It is, as described above, not possible to have totally the same fields of activity due to the different structures and tasks of local public administration in the UK and

- Growth of organisation: number of employees
- Length of implementation: years of implementation of EMAS within the organisation or a part of it
- Depth of implementation: Organisations have usually implemented EMAS in all or a large number of departments.

These criteria were selected to find comparable organisations, regarding the organisational structures as well as a quantitative and qualitative standard regarding the work with EMAS. A third step was to ask the relevant organisations in written form if they would be willing to participate in the study. Some of them did not want to be part of the study so, finally, three UK and three German authorities were selected for the study.

5.3.2 Organisations analysed within this study

5.3.2.1 UK authorities

The following organisations from the UK are analysed in detail within this study:

Authority UK 1: UK 1 is a district council in the West Midlands region. The district was formed on April 1, 1974 in the course of the national local authority's reform by the merger of several smaller districts. It covers an area of 997.87 km^2 and is the 26th largest district in the UK. The authority is responsible for a population of 119,000; the population density is 122 per km^2 (estimated, 2005, information taken from authority's internet site). The authority's headquarters are based in the main town of the area. Up to 2000, the majority was held by the Liberal Democrats, who also introduced EMAS to the authority. The present council majority is Conservative, as is the chief executive of the organisation. The authority started to introduce EMAS in 1996 under a Liberal Democratic majority. This process took about one year; registration was achieved in 1997. The EMAS registration covers the headquarters building. No further sites are included in the environmental management system.

Authority UK 2: The authority is a district council in England's Southern region. Like the district of UK 1, this district was formed on April 1, 1974 by

---

Germany. Nevertheless, there are a number of fields where both UK and German local authorities work in, regardless of the source of finance for these tasks.

the merger of several smaller districts. It covers an area of 292 km² and is the 157th largest district in the UK. The authority is responsible for a population of 93,400; the population density is 320 per km² (estimated, 2005, information taken from authority's website). The authority's headquarters are based in the main town of the area. The present council's majority is Liberal Democrat as is the chief executive. From 1987 to 1991, the majority within the council was held by the Conservatives. In 1991, the Liberal Democrats regained the majority in the council and have held it since then. In the 2007 local elections, where all forty-one seats were up for election, the Liberal Democrats remained the leading party, gaining twenty-three of the forty-one seats. The authority began with EMAS in 1996; first registration was achieved in 1999. In addition to EMAS, the authority gained the ISO 14001 registration in 2002, being the only council in the UK with two types of environmental management systems. The council has been regularly re-validated since the first registration; the next validation was due in 2007.

Authority UK 3: The authority is a district council in England's Southern region. The district was formed on April 1, 1974 by the merger of several smaller districts. It covers an area of 460.65 km² and is the 109th largest district in the UK. The authority is responsible for a population of 110,000 people; the population density is 239 per km² (according to the Office of National Statistics, ONS). It is mainly a rural area with close connections to London. Many people commute by train for some time each day to have the possibility to live in this very scenic region which also attracts large numbers of tourists every year. The authority's headquarters are based in the main town of the area. The present council majority is Conservative as is the chief executive of the organisation. From 1992 to 2000, there was no overall control of the council by a single party. Remarkably, there have been a number of Green councillors in the council for more than a decade. From 2002 onwards, the Conservative party again held the majority in the council. However, they lost it in the 2007 local elections. The organisation started with the EMAS process in about 1996, gaining first validation and registration in 1999. Since then, it has changed from the EMAS I to the EMAS II regulation and has been revalidated regularly. The next re-validation was due in 2007.

5.3.2.2 German authorities

The following organisations from Germany are analysed in detail within this study:

Authority GE 1: The authority is a district council (Kreisverwaltung) in Bavaria, southern Germany. It has about 590 members staff and covers an area of 667 square kilometres. It is responsible for delivering services to 309,080 people around a large city in Bavaria – it covers the rural regions surrounding it. The population density of the region is about 496 inhabitants per square kilometre. Its headquarters are based within this city, which itself is not part of the area the organisation is responsible for. The major political force in the authority is the Christlich-soziale Union (CSU), a regional Bavarian conservative party. The Landkreis has been dominated by the CSU since its beginning. The chief executive is a member of the CSU. EMAS was introduced in 1999; first registration was granted in February 2001. Its management system reflects the EMAS II regulation. The authority has since been revalidated regularly. It has issued the environmental statements according to the EMAS regulation. The revalidation was carried out in April 2007, the next revalidation is due in 2010. The EMAS registration covers the organisation's headquarters which include office space, a canteen, a print shop, several workshops and an underground car park. The authority has a fleet of ten vehicles at this site, mainly for passenger transport (at the time of the interviews). EMAS is not extended to other sites of the authority. However, elements of EMAS, like activities for power saving, are implemented into the management of other sites and buildings.

Authority GE 2: The authority is a district council (Kreisverwaltung) in Bavaria, southern Germany. It has about 270 staff (source: organisation's EMAS environmental statement 2005, p. 4) and covers an area of 1,275 square kilometres. It is responsible for delivering services to 109.219 people in the region (figures from the organisation's homepage December 31, 2005). It is, therefore, among the larger Landkreise in Germany. The major political force in the council is the regional conservative party CSU with twenty-five seats, followed by the labour-orientated SPD with fourteen seats. The CSU has long been the majority party both of the council and of the state of Bavaria. The chief executive is a member of the CSU as well. EMAS was introduced in the council headquarters in 1999; the first registration in the national EMAS register was in 2001. The EMAS system covers the council departments within the main building; other small public organisations that work within the building, like the authority's waste management service, take part in EMAS on a voluntary basis. Other sites or buildings of the authority are not affected by the management system. The latest validation was carried out in 2007; the next re-validation is due in 2010.

Authority GE 3: The authority is a district council (Kreisverwaltung) in Saxonia-Anhaltina, East Germany. At the time of the interviews, it had about 437 members of staff (administration and emergency services, other organisations affiliated with the authority not included) and covered an area of 797,6 square kilometres. It was responsible for delivering services to 91,000 people in the eastern Harz region; it mainly covered a rural area which has a relatively small population – about 116 inhabitants per square kilometre – the German average is 235 inhabitants per square kilometre (source: environmental declaration 2006). EMAS was introduced to members of the authority in 1999. The formal decision to implement it was taken in January 2000. The authority was registered in the national EMAS register in February 2001. Due to a merger of three regional authorities into a new organisation in July 2007 due to the demographic downturn of the region, the current status of EMAS is unclear. It seems that the authority did not continue with it. The major political force in the authority is the conservative CDU party; it supports the chief executive, who is not a member of the party himself. The EMAS registration covers all buildings in the town of the headquarters of GE 3, including the regional ambulance service building and the regional road maintenance facility of the authority.

5.3.2.3 Summary of the samples

The following table summarises the basic facts about the sample regarding the political setting as well as the timeframe of the EMAS processes:

	Authority UK 1	Authority UK 2	Authority UK 3	Authority GE 1	Authority GE 2	Authority GE 3
EMAS process since	1996	1996-1997	1996	1999	1999	1999
Time of implementation	1 year	2-3 years	3 years (1996-1999)	1.5 years	2 years	2 years
Time of registration	1997	1999, ISO in 2002	1999	2001	2001	2001
Leading party at time of decision	Lib. Dem.	Lib. Dem.	Lib. Dem.	Cons.	Cons.	Cons.
Present (2005) leading party	Cons.	Lib. Dem.	Cons.	Cons.	Cons.	Cons.

Table 5.1: Political setting and time frame for EMAS processes of sample organisations. Source: Own.

5.4 The interviewees

The interviewees were chosen in order to get information about the implementation of EMAS within the organisation. Therefore, the aim was to interview the three following levels of hierarchy:

1. Top management, i.e. the organisation leaders (in German local administration: Oberkreisdirektoren or Landräte, chief executive in the UK) who have the overall executive responsibility for the authority and often play a leading role in policy development and strategy formation.
2. Middle management, i.e. the desk officers and EMAS managers that deal with the environmental management system in their daily work, i.e. the environmental managers or EMAS coordinators (in German: Umweltmanager, Umweltmanagement-beauftragter, Umweltmanagementkoordinator etc.).
3. The unit managers leading units or departments with large environmental effects or where EMAS was implemented (if there are units where EMAS was not implemented).

Additionally, some employees working in these departments or ones with experience of EMAS were interviewed, at the suggestion of an authority. Furthermore, there was the chance to interview a member of the council (one present and one former member) at two British organisations that were or are involved with the topic of the study.

The interviews were conducted between April and June 2005. With the exception of one, all were face to face interviews at the relevant site of the authority. Usually, the interviews were single interviews with one interviewee only, as requested in the written information package that each authority received in advance.[75] On two occasions, the authorities permitted group interviews only.[76] At some organisations, it was also possible to take a look at the facilities the interviewees were talking about. The interviews were recorded with a digital recording device[77] as agreed to by the interviewees in

75 This information package included a description of the research project, a reference by the European Commission, DG Environment, a reference of the dissertation advisor and some information about the interview guideline, but not the guideline itself, to prevent deliberate preparation of the interviewees.

76 This was due to the circumstances. Individual interviews were preferred but not always possible.

77 It was an iriver h320 with build-in microphone.

advance. The digital recording was done to facilitate the following data collection and analysis. Furthermore, it helped the interviewer to concentrate on the interview itself. This was very important due to the semi-structure of the interview guideline. Looking at each authority, the interviewees are as follows:

Authority UK 1: All interviews were carried out at the council's building. The six interviewees have the following positions: head of policy and public relations (senior management), EMAS coordinator (middle management), cleaning officer (staff), chief executive (senior management), scientist/ politician (former council leader). One interview was carried out partly with two interviewees (for about thirty minutes, after that, one interviewee left the room).

Authority UK 2: All interviews were carried out at the council's buildings. The six interviewees have the following positions: housing contracts manager (middle management), EMAS coordinator (middle management), principal environmental health officer (middle management), director of planning and environmental services (senior management), councillor (former lead member for environment) and internal auditor (middle management). All interviews were carried out with one interviewee.

Authority UK 3: All interviews were carried out at the council's buildings. The six interviewees have the following positions: strategic director (senior management), facility manager (staff), technical contracts officer for waste (middle management), policy officer, also working on EMAS (middle management), policy officer (middle management). All interviews were carried out with only one interviewee.

Authority GE 1: Four interviews were carried out with staff members of this authority, of which one was a group interview. The director of environmental affairs and planning (senior management) the EMAS coordinator (middle management, took part in group interview first, then a single interview followed) and three office staff were interviewed.

Authority GE 2: The interviews were carried out at the council offices; one was carried out by phone a few weeks later. The interviewees were one senior officer responsible for waste management and environment (senior management), the EMAS coordinator (engineer, middle management, interview by phone), a member of the EMAS team (engineer, middle management) and a caretaker (staff, responsible for building maintenance and technical services).

Authority GE 3: The interviews were carried out as one group interview; the interviewees were the director for administrative affairs (senior management), an EMAS officer (staff) and one person from the staff council.[78]

78 In the course of the interview, at one point the senior member of staff refused to give an answer to a question. It seemed that the interviewees had been prepared by this staff member not to give certain answers.

Organisation		UK 1	UK 2	UK 3	GE 1	GE 2	GE 3
Senior management		Chief executive					
		Head of policy and public relations	Director of planning and env. services	Strategic director	Head of environmental affairs and planning	Head of environmental and waste	Director of administrative affairs **
Middle management		EMAS co-ordinator **	EMAS coordinator	Policy officer	EMAS co-ordinator*	EMAS co-ordinator* **	
			Housing contracts manager	Technical contracts manager		Environmental engineer	
			Principal env. health officer	Policy officer			
			Internal auditor				
Staff		Cleaning officer**		Facilities manager	Caretaker*	Caretaker	EMAS desk officer**
					Procurement officer*		Member of staff council**
					Procurement officer*		
Council members		Scientist/ politician (ex council leader)	Councillor (ex lead member for environment)				
		Councillor (lead member for environment)					
Total		6	6	5	5	4	3

*: group interview, **: partly group interview, ***: interview by phone

Table 5.2: Interviewees of the study and their positions in the organisation. Source: Own.

5.5 Data Collection

The first process to get data for the analysis was the transcription of the digitally recorded interviews using a freely available software tool (Barras et. al. 1998, Barras et. al. 2000).[79] Because the transcription focused on content, standard orthography was used; literary writing was not considered as useful. Nonverbal articulation was only transcribed when it was seen as intentional of the interviewee. If a short answer like "yes" or "no" was given with a particular action, like hesitating or laughing, this was marked as well. The interviews were transcribed in the original language, i.e. in German or in English. About five hundred pages of transcription were produced with this method.

After the transcription of the interviews, the data was collected using the software tool MIA, developed by Gläser/Laudel (2004). This is a macro-orientated tool for qualitative content analysis. Based on VisualBasic, it enables a content analysis where it is possible to use the attributes of the empirical material as free verbal descriptions (Gläser/Laudel 1999, Gläser/Laudel 2004) whereby only the variables defined by theory are fixed within an ex ante process; their characteristics are developed within the process of the analysis (Gläser/Laudel 1999, p. 10). This method leads to a structured and condensed result of the analysis. Compared to the empirical material of the interview transcriptions, the analysed data is reduced to a manageable amount and can be used for further analysis and interpretation. The method was adapted to the specific situation of this research project by creating and defining the macros which were needed for the analysis. The method was chosen to cope with the large amount of text that derived from the transcription of the interviews. The data analysis could also have been done with the traditional coding method but this was neglected because it has been found that it is orientated on quantitative ideas of analysis, not on qualitative ones. And what is more, the traditional coding seems quite static as it follows a defined and closed system of categories (Mayring 2003), whereas the chosen method is more flexible to be adapted to the needs of the analysis during the data generation process.

The process of generating data using this method is called 'extraction'. Wheras traditional coding indexes a text so that both the text as well as the index are the source of the analysis, the extraction method takes information out of the

---

79 It is the software tool 'transcriber', developed by computer linguists, available at http://trans.sourceforge.net

text and structures it. This condensed information is the basis for the analysis to follow (Gläser/Laudel 2004). These extracted texts (or tables) contain only information that is relevant for the research question. The categories that are used during the extraction process derive from the theoretical model of chapter four. They are collected in nominal scales.

The whole process of the qualitative content analysis consists of the following four steps: preparation of extraction, extraction, preparation of data and, finally, data analysis. The methodological preparation of extraction covers the construction and validation of the variables which are checked against the theoretical model. Thus, it can be evaluated whether a variable that is important within the theory was irrelevant for the interviewees or whether there is a need for further variables (Gläser/Laudel 2004). The technical preparation was to construct the extraction modules for the categories with the MIA software, which is a tool to construct and use extraction macros (developed by Gläser/Laudel (2004)). For each variable, a macro was constructed which contains the forms of the variable. The construction of the variables was orientated towards the theoretical model laid town in chapter four. The elements of the theory with their specific issues have been taken up in the design of the variables but the policy cycle is dominating the design of the macros which form the matrix for the extraction process.

Policy Process		
Phase	**Variable/Question**	**Form**
Problem Perception	Acteurs/Who?	Executive
		Environmental Management Coordinator
		other member of staff
		Ministeries
		authorities on a higher level
		other authorities
		Environmental organisations
	Time/When?	given in interview
	Media/How?	Conferences
		Brochures
		through staff
		on request
		Others
	Reasons/Why?	environmental concern/problems
		Image
		pilot project
		public authority reform
		concrete problems at organisation

		on request
		Others
Agenda setting	Acteurs/Who?	Executive
		Environmental Management Coordinator
		other member of staff
		Ministeries
		authorities on a higher level
		lobby groups
		Parties
		environmental organisations
		Other
	Time//When?	given in interview
	Reasons/Why?	Image
		environmental safety
		pilot project
		public authority reform
		energy saving programme
		performance indicators
		Other
	Procedure/How?	concept paper
		working group by officers
		working group by legislative body
		working group by party
		Other
Policy Formulation	Acteurs/Who?	Executive
		Environmental Management Coordinator
		other members of staff
		working group by officers
		working group by legislative body
		Ministeries
		other authorities
		Consultancy
		Others
	Time/When?	given in interview
	Reasons/Arguments/ Why?	environmental health
		pilot project
		public authority reform
		energy saving programme
		improvement of management procedures
		efficiency (finances)
		modernising of infrastructure
		environmental programme /EMAS requirement
		Other

	Procedure/How?	own procedure
		EMAS guidelines
		with consultants
		help from other organisations
		Consultancy
Decision making	Acteurs/Who?	Executive
		legislative body
		executive body
		environmental department
		higher authority
		Other
	Reasons/Arguments/ Why?	problem perception and agenda setting
		Other
	Time/When?	given in interview
	Procedure/How?	top-down
		bottom-up
		consensual decision
		Other
Policy implementation	Acteurs/Who?	Environmental Management Coordinator
		working group for whole organisation
		working group for unit
		environmental department
		Consultants
		Auditor
		Other
	Reasons/Arguments/ Why?	see Problem perception and Agenda Setting
		Other
	Units of implementation/Wher e?	whole organisation
		whole site
		single unit/units
		Other
	Time/When?	given in interview
	Direct environmental impacts/What?	Electricity
		Water
		gas/energy
		Stationery
		Fuel
		other (specify)

	Indirect environmental impacts?	planning
		Housing
		Procurement
		land use
		Other
	Sucess examples/What?	see evaluation
	Problematic examples/What?	see evaluation
Policy Evaluation	Acteurs/Who?	Environmental Management Coordinator
		internal auditor
		external auditor (validation)
		team leader
		working group
		higher authority
		Consultants
		Other
	Procedure/How?	internal audit
		external audit/validation
		environmental programme /EMAS requirement
		according to EMAS procedures
		Other
	Sucess examples/What?	Savings
		energy savings
		water savings
		stationery savings
		gas/energy savings
		other savings (monetary)
		better performance (examples)
		performance indicators
		legal certainty
		better image
		Other
	Problematic examples/What?	slow implementation
		large paperwork
		staff did not cooperate
		staff needed to be convinced
		technical problems
		financial problems
		problems of bureaucracy
		change of staff

		lack of support from top management
		lack of support from other units
		Other
Policy Termination/restate of cycle	Acteurs/Who?	Executive
		Environmental Management Coordinator
		other member of staff
		Ministeries
		authorities on a higher level
		other authorities
		legislative body
		executive body
		environmental organisations
		Other
	Reasons/Arguments/ Why to continue?	better image
		improvement of management procedures
		improvement of performance
		procedures are integrated
		expectation by higher authority
		expectation by customers/the public
		convinced by EMAS
		more problems came up
		other (specify)
	Reasons/Arguments/ Why discontinue?	Costs
		lack of staff
		other political priorities
		procedures are integrated
		change of political will
		other (specify)

Table 5.3: Variables deriving out of the theoretical model and used for the analysis. Source: Own.

The extraction was done by reading the texts. For each paragraph, a decision was taken for which variables it contains information. For thematic and methodical reasons, no rules of extraction were defined in uncertain cases. There are two reasons for this. First, the variables largely follow the structure of the policy cycle, which itself is the basis for the interview guideline. Second, if there was any doubt regarding allocation to a variable, the information was allocated to both variables and this was marked. This was done to maintain the connection of content between both variables.

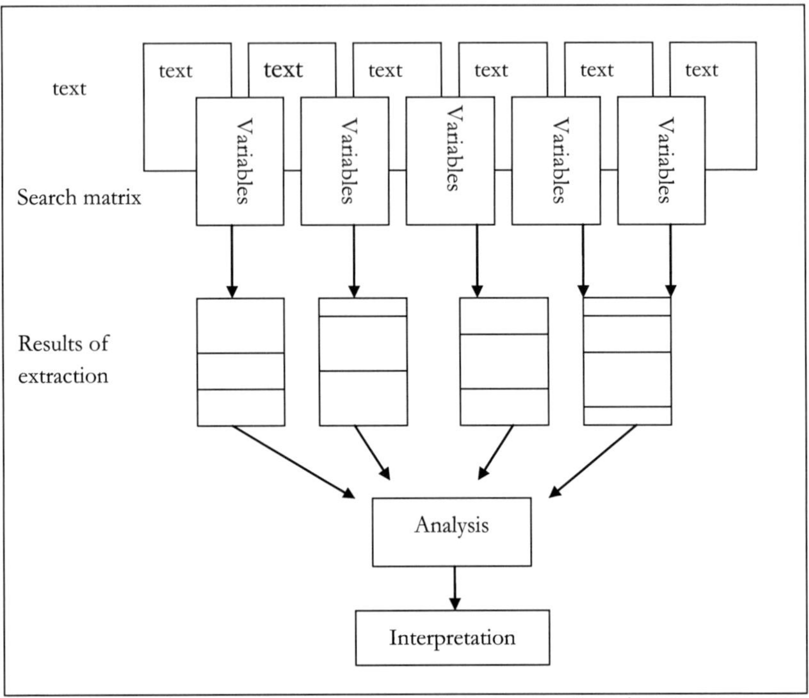

Figure 5.2: Work flow of qualitative content analysis. Source: Gläser/Laudel 2003, p. 194.

With some variables, forms were added throughout the extraction process. The extraction process was a long one due to the number pages of the transcriptions. The result of the extraction was a large set of tables that contained the variables and its forms. This data set has about two hundred pages.

As a last step before the analysis, the data was checked and prepared for the analysis. This was done through the grouping of information that belonged together but was placed at several variables, a round-up of information that had the same content and a correction of visible mistakes.

One central advantage of the extraction is that it can be adapted to new knowledge in the course of the extraction process. An important element for the reliability of the extraction is the verification of data. This is secured with the help of the software tool MIA, which contains a reference to the original data for each finding. Through the whole process, a data set is generated that categorises the information and contains a reference of the original source.

5.6 Data analysis and interpretation

The result of the data collection was a structured set of tables that contained the information of the interviews in a summarised form. These were the basis for the interpretation of the data as set out in chapter six. First, an analysis is given for each authority. It is done on the basis of the three theoretical models (concept of public authority, policy cycle and organisational learning). The process of introducing and implementing EMAS is analysed, followed by an analysis of the learning outcomes of these processes. At the end of the chapter, an overall conclusion to the EMAS processes is given for each authority. Finally, general conclusions regarding EMAS in public authorities are drawn as a result of the analysis. These are laid down in chapter seven. Here, general comparisons between the situation in the UK and Germany are made.

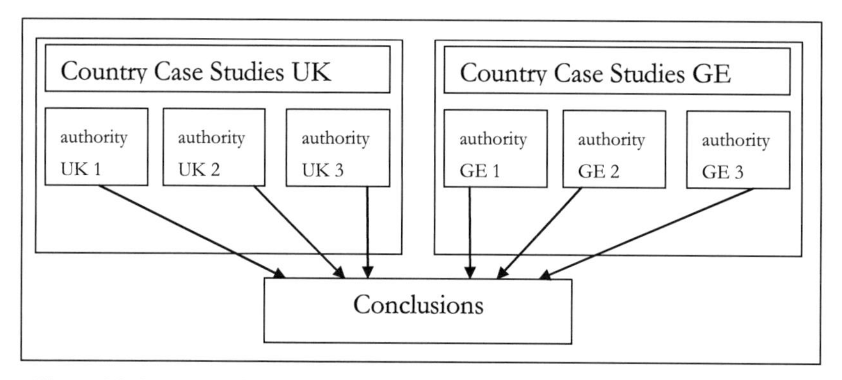

Figure 5.3: Structure of data analysis and interpretation. Source: Own.

As a result of the empirical work, a detailed view of the introduction and implementation of EMAS in local public administration in the UK and Germany is given.

6 Empirical Findings

6.1 Introduction

Within this chapter, the main aim is to describe and analyse the outcomes of the interview analysis as methodologically described in chapter five. To begin with, in the following, the main questions of the analysis will be recapitulated in a comprehensive form.

First, the two concepts, the classical Weberian orientation of an authority or the modern, market-focused way of working under the NPM model describe the two antipodes of today's public authority management. Within this analytical part, it is examined whether the authorities work more towards one of the two models and whether the management contains elements of both models. The variables used here are based on the ones given in table 4.2. The findings of this part of the analysis give an indication of the management culture of the authorities and thus answer the research question of the effects of the management culture on the implementation of EMAS as a policy instrument.

Second, the introduction and implementation of EMAS is analysed with the policy cycle, keeping in mind the different variables set out in table 5.3. The following questions lead the analysis of the interviews in this part: for the problem perception phase, the question of discrepancy is the leading one. Within the agenda setting, there is the question of whether the process can be described as a policy adoption or a policy innovation and whether this process was initiated by society or by a public authority. The policy formulation phase concentrates on the actors involved as well as any other important information that is of relevance for this phase. Within the decision making process, the focus is on the actors who made formal as well as preliminary decisions. A second question is whether the process was a closed one or one involving a plurality of actors. The policy implementation phase investigates the responsible actors, takes a look at the allocation of resources, the actors who work with the policy programme and examines special cases within this phase of the policy programme. Further, the different actor types are examined here – programme conceptionalists, implementations managers, street bureaucrats as well as the addressees of the programme. In addition, the influence of this policy programme on other programmes is looked at. The policy evaluation phase concentrates on the methods and means of evaluation as well as the question of policy learning. Finally, the policy termination phase investigates reasons for termination or continuation of the policy programme. These variables give a more detailed, yet consecutive view of the different processes and actors within the phases. The findings of this part of the analysis will answer the question of how the policy instrument EMAS was introduced and implemented within the organisations.

Third, the findings of the parts of the interview concentrating on evaluation of the policy programme are the basis for the analysis of the organisational learning processes, which are based on the theoretical model of organisational learning outlined in chapter five. The results of both double-loop learning as well as deuteron-learning are structured along subcategories in order to achieve a better comparability of the analysis' results. These subcategories were developed after the content had been analysed through the policy cycle model; they derive directly from this phase of the analysis. The results provide valuable material for the learning processes of the organisation. They answer the question of whether or not learning processes took place while implementing EMAS. Furthermore, within this chapter, it is examined whether the continuous approach to change can be applied within public authorities. Table 6.1 summarises the structure of the analysis in reference to the theoretical model used here and gives an indication on the expected outcomes.

Concept	Phase	Question	Outcome
Classical Weberian operation of public authority	Not applicable	What is the underlying mode of operation? More classical Weberian or more NPM-orientated	Indication of management culture of organisation
NPM-orientated operation of public authority	Not applicable		
	Policy cycle		
	Perception Phase	Is there a discrepancy between reality and policy perception?	Introduction of the topic/the policy instrument to the organisation. Political setting and political priorities of the time
	Agenda Setting Phase	Policy innovation or policy adoption? Who was the initiator in this phase?	Mode of organised policy innovation, importance of actors within this phase
	Policy Formulation	Who are the relevant actors? Additional information?	Mode of operation of relevant actors

	Decision Making Phase	Actors who made formal decisions? Actors who made preliminary decisions? Closed or plurality decision?	Mode of decision-taking, including formal and preliminary decisions. Mode of decision-making culture
	Implemen-tation Phase	Who are the relevant actors? Which actor types are relevant? What about the allocation of resources? Influence on other policy programmes?	Activity of actors within this phase. Importance of leading actors. Allocation of resources and influence on other policy programmes as well as on the operation of the organisation
	Evaluation Phase	What kinds of evaluation? What kinds of policy learning?	Ways of evaluation of the policy instrument and its effects
	Termination Phase	Why does an organisation terminate or continue with EMAS?	Reasons for continuing or terminating the EMAS policy instrument. Influence of other decisions on EMAS?
Organisational learning			
	Espoused theories	How does the organisation want to present itself?	External image of the organisation
	Theories-in-use	How does the organisation operate in reality?	Internal way of communicating and operating EMAS. Discrepancy between espoused theories and theories-in-use.
	Single Loop Learning	What kind of small-scale improvements have been achieved with EMAS?	Technical improvements, small changes due to EMAS
	Double Loop Learning	What kind of operational (larger) improvements have been achieved with EMAS?	Operational improvements due to EMAS. Subcategories: database issues, communication work with the regulation other activities

	Deuteron Learning	What kind of organisational, overall changed processes took place due to EMAS?	Organisational improvements due to EMAS. Subcategories: overall learning management system follow-up projects other relevant aspects
	Negative Issues of the organisational learning[80]	What kind of negative processes accompanied the implementation and use of EMAS?	Discussion of negative issues of the EMAS processes

Table 6.1: Structure of analysis in reference to the theoretical concept. Source: Own.

The results of the analysis are presented for each organisation. The analysis begins with authorities in the UK followed by the German authorities. To underline the findings, some quotations of the original data are cited. The German ones have been translated by the author.

6.2 Authorities of the United Kingdom

6.2.1 Authority UK 1

6.2.1.1 Concept of public authority

When looking at this authority from the viewpoint of the underlying concept of public authority, it is clear from the evidence of the interviews as well as from the environmental statements that the aim of the organisation is to be modern and efficient. The structural guiding principle is to have a best value performance outcome and to deliver the best services to the people. The organisation is outcome orientated; its activity guiding principle is adaptation. The organisation is focused on final programming; the mode of work and the documentation needs to resemble the principles of transparency, efficiency and the national key performance indicators. When problems occur, the

80 This aspect was not part of the theoretical framework. In the course of the data analysis, it became evident that numerous interviewees were not only describing positive effects of organisational changes in connection with EMAS but also negative ones. Therefore, these aspects are also added to the analysis to get a more complete picture of the processes that took place.

termination of the business and/or a legal process can be assumed. Regarding EMAS, the officers show partly intrinsic motivation and are part of a relatively flat hierarchy. To summarise, the authority is a classical case of a NPM-orientated authority.

6.2.1.2 Policy cycle of EMAS introduction and implementation

Within the *policy perception phase*, the emerging of the Green Party in the UK as well as the discussion about environmental issues that came up in the early 1990ies made it clear that there was a discrepancy between the actual environmental policy of the organisation at that time and the need for a new, more environmentally focused policy and an overall manage-ment of its activities. This was the reason for the authority to work out a strategy to imple-ment EMAS. Furthermore, the Liberal Democrats had a majority within the council in 1997. This party was and still is in parts much more environmentally aware than the Conservatives or the Labour Party. The council leader at that time was a scientist who was environmentally aware and had come to know the EMAS standard through its personal professional background. In addition, there was an environmentally aware senior management leading the authority. These actors were the leading figures for the perception of EMAS as a tool to manage the environmental impacts of the organisation.

Regarding the *agenda setting phase*, EMAS was the first management system for the authority. The process of agenda setting can be described as a policy adoption because the relevant actors took the EU regulation as a framework to develop and implement their own environmental management system. Nevertheless, as there were not many other organisations from which one could get information about how to work with EMAS and due to the fact that there was almost no support from the national government, the process had, of course, innovative elements for the organisation. Over the last years, the environmental management system has been integrated in an overall general management framework for the whole organisation. This development was encouraged by the national standards that were imposed upon the local authorities in the local government reform (see chapter two for details).

The *policy formulation* was done by the later EMAS coordinator, a working group of officers, some councillors as well as the council leader. The different actors involved in this process showed a large interest and commitment towards this policy programme because at the time of the policy formulation, environmental activities were very important for them. Additionally, the interviewees reflected on the importance of the environmental activities and the support the EMAS implementation process received from all relevant groups of internal and external actors in the mid 1990ies. Thus, the work with EMAS can be seen as a reaction to a general political trend to consider environmental issues that were strong in the region.

The *formal decision* to implement EMAS was taken by the council under its Liberal Democratic majority in 1996. Preliminary decisions had been made by the EMAS coordinator, the EMAS working group and the council committee on environmental health. The interviewees indicated no clear evidence for the closeness or the plurality of the decision making process, although the number of actors as well as the political majority at the time of decision making suggest that the decision process tended to be pluralistic rather than closed.

The *policy implementation*, i.e. the introduction and implementation of the various steps of the EMAS system, was led by an enthusiastic full-time EMAS coordinator. In 2001, when this person left the organisation, the post was reduced to a part-time position and filled with a more conservative officer in 2002. The main work with the overall programme is done by an officer who has the function of the EMAS-coordinator, while the individual officers work according to the rules and standards of the system. The allocation of resources is determined by the council and the chief executive. These make their decisions based on the current rules set by central government which gives more than eighty per cent of the financial resources directly to the council. Only about twenty per cent of the revenues are sourced from the region.

Over the years, the council has been quite successful in its environmental activities and won a national recognised price for them in 1996. The policy programme of EMAS was drawn up by the actors involved in the formulation of the environmental policy and programme under the EMAS regulation. The implementation was led by the EMAS coordinator. The other officers of the different departments work in accordance with the rules and procedures within their standard routines. These staff can be described as the street bureaucrats according to the theory. The addressees of the programme are all staff as well as the general public affected by the work of the council together with the contractors who deliver services for the council.

The environmental programme and the environmental policy have definitely had an influence on other programmes, like contracting, procurement, waste collection and the management of council housing because all these programmes have an environmental impact. Nevertheless, within its environmental policy, the council primarily focuses on direct environmental impacts. Indirect impacts are rarely tackled. One example was the procurement of vans running on LPG rather than on petrol.

The *policy evaluation* is carried out according to the EMAS regulation with both ex post as well as ex ante evaluation on a regular basis. Additionally, reporting to the general public with the environmental statement is a necessary requirement. Therefore, the environmental activities under EMAS are one of the best evaluated policies of the council. Nevertheless, within the national key performance indicators, the EMAS policy programme is not taken into account and, consequently, there has been a decrease of activity and interest in it because under the national management standards ('best value performance'), councils are primarily interested in getting a good record of the management standards that are regularly checked by the Audit Commission of Central Government.

Up to now, the authority has continued to operate EMAS as its environmental management system and (at the time of the interviews in April 2005) expressed its commitment to continue with it. The next validation was due in 2007. There is no current information on the status of the validation (as of January 2009).

6.2.1.3 Organisational learning

Regarding the difference between espoused theories versus theories-in-use, one can say that the organisation is environmentally aware – although importance and awareness levels have changed over the years. One major factor for these developments is a change of the political majority within the council from the Liberal Democrats to the Conservative party as well as a general decrease of the interest in environmental policy up to the middle of the 21st century when the climate change brought the topic back on the agenda.

Within the espoused theory, the interviewees expressed their view that the authority is able to produce a better image to the general public through its EMS. In addition, improved communication within the organisation, the active delivery of community responsibility and leadership within the field of environmental policy as well as the active communication strategy through the environmental statement are effects of the espoused theory. The underlying theories-in-use show quite a different picture. According to the interviewees, the most important thing is to keep the system and not spend too much on it in order not to lose their positive image or the European registration. This was due to the fact that the general interest in the system had decreased:

(Voice000-1-22): "So when it first started, lots of interest, lots of involvement, everybody across the council doing many things, and so, big improvements in environmental performance over the first five years or so. It gets hard to keep that interest up, so, over the last few years, the interest has dropped because we have achieved all the easy wins."

Despite the efforts for a good communication about the environmental activities, the organisation got largely negative or no responses at all from local businesses or the general public about their EMAS activities because the general public does not seem very interested in the internal activities of the council; only external activities are recognised and regarded as important. This is due to the outcome orientation where only figures and services count. Thus, the procedures determining how these outcomes are generated are not of public interest.

When looking closely at the different levels of learning, one can see that within the *single loop learning*, a lot of technical improvements have been made to reduce the direct environmental impact of the organisation, i.e. changing procedures of operation, documenting the use of energy, water, waste, etc. to see if there are any peak times of use in order to reduce these. These direct environmental impacts are the major focus of the organisation's work with EMAS.

Within the *double loop learning* phase, four different clusters can be identified: database issues, communication, work with the regulation and other activities.

First, database issues of the environmental inventory can be summarised by the systematic establishment and maintenance of the relevant data set that is needed for the management system according to the EMAS regulation, such as the use of gas, electricity, water, petrol, stationery, the register of laws that are important in the field of environmental policy, the use of machines, vans, etc. Prior to EMAS, there was more or less no such data available. These aspects were not regularly monitored or not monitored at all. When EMAS was integrated into a more general management system, partly due to the

KPIs, the data collection of the council gained a broader focus and was developed more systematically.

Second, communication structures have changed due to EMAS. Before the introduction of the management system, there was no systematic communication about environmental issues. EMAS established a constant system of checks as well as a regular reporting every six months. Through this, the heads of department, the chief executive as well as the council and its committees are permanently aware of the achievements of the organisation regarding environmental issues. If targets are not met within the timeframe that has been set by the executive and documented in the environmental policy and management plan, other methods and procedures that are considered to be more effective are quickly put into place. Another important issue that can be described as double-loop learning is that at the beginning of the work with the regulation, it was a special driver for changes within a specific field of the council's environmental policy, namely its refuse and recycling activities. The organisation realised that these had large impacts on other fields of activity as well as on all people living in the district. Thus, numerous activities were conducted to reduce the amount of waste produced within the district.

The *deuteron learning* phase is the most important one for a successful implementation and maintenance of EMAS because it demands the most comprehensive changes of the tasks and the general strategy of the organisation. Here, the outcomes of the analysis are clustered again to get to a more consistent picture. This time, the four clusters are 'overall learning', 'management system', 'follow-up projects' and 'other relevant aspects'.

First of all, there are overall learning outcomes that have been achieved by the organisation. The council has developed methods on how to work with a quality control management system which needs data and key figures to be administered. EMAS was the council's first management system and also the first time that an external person (the verifier) would look closely at the activities of the authority and evaluate these:

(Voice0001chris-146): "If we just not had done EMAS, we wouldn't be checking it as much as we are to actually see what the effects are."

As part of the EMAS idea, the organisation has to commit itself to a constant improvement process which implies reflection on the past activities and a strategic planning for future policies. As a consequence, the environmental management system has had an impact on the whole organisation by producing a change of thought more than anything towards a sustainable and long term perspective.

Regarding the management system itself, the organisation has learned how to interpret the European EMAS regulation and subsequent documents over the years. From a strict implementation at the beginning, taking the regulation word for word, the organisation has acquired a more flexible and demand-orientated approach. After an intrinsically motivated start of the policy programme in 1996, the integration into a broader management system in 2001 together with staff changes and a certain loss of intrinsic motivation made it more difficult to operate EMAS. When the organisation took EMAS into its focus again in 2002, the structures were streamlined and the procedures were changed to make it easier for officers to follow them while not losing the essence of EMAS. Furthermore, the widespread approach to tackle all environmental outcomes of the organisation has been reduced within this re-orientation to tackling only the most important issues. This development of the policy programme indicates the changed political commitment and support over the years. During this change of attitude towards the management system, a yearly check on costs and benefits of the system has been put in place. Through the NPM-orientation of the organisation, this "value for money" orientation has been introduced for almost all policy fields, leading to a large re-organisation of tasks and structures.

A second major cluster of findings are possible follow-up projects that derived out of the activities around EMAS. With regard to the council, a large survey of what kind of services and activities people expected from their council – not only in terms of the environment – was conducted, analysed and put on the political agenda. Furthermore, some aspects of energy management as well as the reduction of energy consumption that had been introduced in the council's main buildings have been applied to the council's leisure complexes. Within the last years, some of these buildings have been transferred to the county council. Alternatively, the management of other leisure complexes has been given to private contractors. Consequently, the council has lost its direct influence on management regarding EMAS issues.

6.2.1.4 Negative issues of the learning process

The interviewees of this council also expressed negative aspects of the process of working and learning with EMAS which are as important as the positive developments in order to understand the whole policy process. When the organisation started with EMAS in 1996, the intrinsic motivation, the idea of doing everything well, better or even best, made the management system very large and powerful. This development was described as negative by the interviewees saying that there was the feeling that EMAS drove the council, not the other way round.

(VOICE000-1-20): "And there was a feeling I think of some people that EMAS was driving the council, not the council driving EMAS."

Another negative effect was the fact that some officers thought the former EMAS officer had been a political lightweight. Consequently, the policy programme was regarded as less important by some staff members. This situation changed over the years, especially in 2001 when the system lost track due to a political change of majorities in the council from the Liberal Democrats to the Conservatives. This also resulted in a change of staff; the people who had begun with EMAS left the organisation. At that time, the organisation realised that the environmental management system needed to be integrated into a larger concept of a more general management system of the council in order to survive in this setting. Furthermore, many of the issues tackled under EMAS would have an effect on other policies. Therefore, the integration was absolutely necessary.

Other problems of the EMAS' implementation were the demand of the verifier to work on a whole range of things whereas the organisation tended to concentrate more on a few aspects over the years as well as the lack of response to the EMAS activities by the public, central government or business organisations of the region. Most respondents to EMAS came from scientific communities and from other authorities, nationally and internationally.

With the introduction of nationwide auditing in the 1990ies as well as the 'key performance indicators' and the 'value for money' and the later 'best value' programme (see chapter two for details), the problem came up that although the council invested resources in EMAS, these investments were not recognised by the standards set up by the national government.

(Voice000-1-92): "EMAS doesn't feature anywhere in how we are audited by central government."

Consequently, under the national rules for performance, the EMAS procedures were additional work (or an additional policy instrument) and not integrated into other existing policies. Thus, interviewees complained about the situation and pointed out that the lack of recognition of EMAS by central government has had a direct effect on the number of authorities using the EMS standard:

(Voice000-1-92): "If part of the regulations said you can demonstrate what central government wants by being registered to EMAS, that would influence the number of authorities that did it."

Nevertheless, the lessons learned from the project, the outcomes generated with it and the capabilities developed within the process have their part in the performance outcomes of the organisation.

Despite the fact that the costs of the management system are under review every year, there are officers that admit that the council has never really done a cost-benefit analysis of the system. Finally, some staff members describe the system over the years as declining, in danger and in their words "hollow":

(Voice008-44): "Yes, essentially, we are a green council, we are EMAS-registered, we are keeping our registration, but I think it is hollow. I think it is a situation whereby we do not give it serious attention. We do not permit environmental issues to impact on our decision making. Our decision making is all about pounds, value for money, not environmental benefit."

6.2.1.5 Overall conclusion of the implementation process within this council

Compared to other authorities on a European level, this council was an early one working on EMAS. The policy programme was begun in 1996; the successful validation and registration was gained relatively quickly, in 1997. As EMAS has now been worked on for a decade, one can say that within the first years, up to 2001, the system was built up and had numerous positive effects on the environmental policy of the authority. When the political setting changed in 2001 from a Liberal towards a Conservative council majority, the activities in the system declined. This also affected its outcomes. Overall, the pressures from central government to focus on the national target setting, based on the 'best value' principle and focussing on the financial side of service delivery while at the same time neglecting the efforts and achievements gained with the environmental management system, made it more difficult for the council to continue. Together with the general lack of interest and the integration of EMAS in the overall management system, it became less prominent over time. EMAS has led to organisational learning through several activities. Nevertheless, there is an obvious discrepancy between the espoused theories and the theories in use; the organisation has a different view of

EMAS than the general public. Additionally, it wants to be environmentally active but at the same time is not able to act more freely from the national performance indicators that strongly determine the authority's activities. At the time of the interviews, the council was committed to continuing with EMAS. However, there is no information available about the status of their environmental management system in the first quarter of 2009. While the organisation is still documented in an internal list of the EMAS help desk, no information is publicly available through the national competent body of the UK. The EMAS revision was due in 2007.

6.2.2 Authority UK 2

6.2.2.1 Concept of public authority

The aim of the authority is to be a modern and efficient service provider for the local people, delivering services to meet the people's needs according to the 'best value' principle. The organisation is outcome orientated and works according to the adaptive activity guiding principle. The type of programming is final while the (mode of) work and documentation is done according to the principles of transparency, efficiency and key performance indicators. Problems are solved by the termination of business and/or a legal process. The attitude of the officers is a partly intrinsic motivation regarding environmental activities; many officers are highly motivated. The organisation has a flat hierarchy.

6.2.2.2 Policy cycle of EMAS introduction and implementation

The *policy perception process* of EMAS can be described as a discrepancy between the arising political and societal needs in the field of environmental policy and the capabilities and activities of the authority in the mid 1990ies. In 1996, the organisation began to work on EMAS, when the internal Local Agenda 21 team started to search for instruments to improve the internal management structures. Furthermore, there was an intrinsic motivation of some staff as well as the political will of the council's majority to do something:

(Voice27.done-12): "I think it was strong environmental commitment by a number of senior councillors and a number of enthusiastic staff. The two things came together. The political group running the council then - it is the same group now, the Liberal Democrats - were very enthusiastic about environmental management. And there were a number of staff here who were very enthusiastic. And we saw EMAS as the best accreditation to go for."

Heavy floods within the region in the year 2000 increased the environmental activities of the authority. In addition, as people were personally affected by the floods, this incident had a strong impact on the environmental awareness of a large part of the population.

The *agenda setting* can be described as an adoption of a policy programme, although there was very little information and support available on how to work on EMAS at the time when the authority began with it. Therefore, the introduction and implementation process was partly innovative because the authority had to develop the concrete elements of the management system according to their needs without any guidance.

Within the *policy formulation phase*, the organisation's Local Agenda 21 group together with the later EMAS officer worked out the environmental programme, the policy as well as strategies for working with the regulation. The starting point for all activities was the council's recycling scheme that included the establishment of a regional recycling centre as well as numerous activities to reduce the amount of waste in the region. The council also combined its social policy with the recycling scheme by providing new jobs for long-term unemployed people in the recycling centre.

The *decision* to implement EMAS was made by the council as well as the chief executive. Preliminary decisions were made by the working group. The decision process was – in contrast to other authorities of the study – predominated by a large amount of plurality and cooperation because the initial idea came from within the organisation and was largely supported by staff.

The *policy implementation* was led by a full-time environmental officer who is responsible for the management system. This officer works in the organisation's recycling centre, which is the leading executive unit for all environmental issues. The other officers of the organisation work with it wherever the regulation has an impact on their daily work. The chief executive is responsible for the allocation of resources that are needed for the system. Through the work with EMAS, the organisation took part in numerous other EU-founded projects, such as the ECObudget project[81], as well as promoting environmental activities for local authorities on a European basis. Further, the organisation established contacts with numerous cities and councils who were

81 See ICLEI's ecobudget homepage for further details on the project of saving resources for local authorities (http://www.iclei-europe.org/index.php?id=1769).

interested in the topic of ecologically-orientated organisational management. Consequently, the organisation has gained a very prominent and positive image among local authorities that are environmentally aware. Regarding the groups of actors that dominated this phase, the chief executive, the council and especially the council committee on environment were active when working on the programme concept. The implementation was led by the EMAS officer while the other officers worked with it through their routine work, thus fulfilling the role of the street bureaucrats. The addressees of the programme were all staff members, the general public, and contractors of the council as well as regional companies who had an interest in environmental activities. The use of the environmental management system has had enormous impact on the following other policy programmes: contracting, procurement, waste collection, energy procurement and consumption, employment of long-term unemployed people, as well as the management of council housing.

As the policy instrument requires both ex post as well as ex ante *evaluation of the programme*, including external verification as well as a public statement, it is a comprehensively-evaluated programme. The activities of the council include the work of the organisation's own internal audit specialist as well as a questionnaire that was sent out to the general public to get to know more about the preferred needs and services, not only regarding environmental issues. Overall, one can see that the organisation has a sustainable, long-term policy approach for tackling environmental issues within its area of responsibility.

(Voice25.done-184): "A lot of that work probably would not happen if we did not have EMAS. Or it may happen and finish. You know, we all need a thing keeping and driving the momentum. And we have that. I think, probably as I said, one of the most important things is the culture. But we have a process so that we include environmental procedures into all our decision-making. And I think that's very important, as well."

In addition to EMAS, the organisation is registered under ISO 14001, primarily because the organisation realised that EMAS is in large parts equivalent to ISO 14001.

(Voice25.done-158): "Well, because in 2002 we thought we are doing ISO 14001 because we have EMAS. So we might as well have a batch. So we decided to get credited for both."

Up to the time of the interviews (May 2005), the organisation had no intention to *terminate* the policy programme; it wanted to continue with EMAS and has done so to the present time.

6.2.2.3 Organisational learning

The organisation's espoused theories represent an image of an environmentally-aware organisation. It has established a mission statement with five core values; one of them is saving the environment. The organisation's main fields of environmental activity are waste reduction, recycling and the reduction of energy consumption.

The theories in use resemble an environmentally-active organisation that has inherited environmental issues. EMAS is a comprehensive and consistent approach to managing the environmental outcomes of the organisation:

(Voice026.done-14): "[...] if you got fragmented approach you don't end up with a very consistent authority."

The authority actively endeavours and communicates its activities. Although the environmental statement is widely distributed and often downloaded, there are no detailed numbers of these downloads and mailing activities. Within the interviews, staff as well as management expressed a commitment to work on EMAS-related projects because this is part of the organisation's mission statement as well as of the organisation's culture. The authority has achieved a consistent picture of both espoused theories and theories in use, at least regarding environmental issues.

Within the *single loop learning* process, a lot of technical improvements as well as changed procedures for the use and management of waste, water and energy use have been implemented. Activities regarding the direct environmental aspects of the authority are in the focus of the management system.

The *double loop learning* is again structured according to the four subcategories: database issues, communication, working with the regulation and other relevant issues. Among the database issues, the council established a database for the monitoring of energy consumption through staff that was responsible in each building. The communication about environmental issues was already on a high level within this organisation. Consequently, the environmental programme and policy needed only minor changes when EMAS was implemented in order to meet the requirements for the environmental management system. It is noteworthy that the organisation has established a culture of improving and learning, where staff is motivated and encouraged to do so. Through the work with the regulation, the need for a corporate approach towards EMAS became evident. Furthermore, the audits revealed both the need to focus on the most important issues as well as the need to be realistic in the goals to be achieved. The interviewees solely remarked that it would be easier to work on external environmental activities, like the vehicles depot or the waste management, compared to office buildings because with the latter, some reductions are only to be achieved through a changed behaviour of the users which is a long-term task but relatively cheap. Other savings can be achieved through technical improvements which are expensive and sometimes take a long time to come into effect.

The outcomes of the deuteron learning process are again subcategorised with the four categories 'overall learning', 'management system', 'follow-up projects' and 'other relevant issues'. This phase of the learning process and its analysis revealed that the organisation has learned that environmental management systems are a good tool to start reflecting on, improving and changing its environmental behaviour, impact and outcome. Furthermore, the implementation revealed that the organisation already had a good environmental policy in place because there were only minor changes necessary to comply with EMAS standards. Through the auditing cycles, it became clear that it would be hard to leave out problematic issues. Consequently, the organisation was faced with tackling these problematic which would have otherwise probably been neglected.

(Voice026.done-46): "And the benefit of this organisation for EMAS means that you are automatically organised for other things, as well. So, there is a strong level of coherence in what the organisation is trying to do."

In the national corporate performance assessment, the council got very good results during the last years due to the capabilities of transferring the knowledge of management techniques onto other policy programmes. Additionally, the use of the environmental management system has changed the way the organisation works with its own property: from just paying the bills towards looking at what is really happening there and what needs to be changed in order to reduce costs and environmental impact.

Regarding the follow-up projects that were generated out of the EMAS-process, the organisation has developed a wide range of activities. One is to provide support to SMEs if they want to improve their environmental impact or even introduce an environmental management system. Another is the active involvement of the organisation in European projects such as ECObudget. Additionally, the council has been the host for numerous groups from councils and public authorities from all over Europe that were interested in environmental management. These follow-up projects indicate that the council has been building upon the knowledge gained with EMAS to deepen and strengthen its environmental policy as well as its other capabilities. Another important factor is that EMAS has strengthened the long-term perspective in many policy areas rather than taking a short-term approach. It has become clear that many of the activities done by the council only make sense in the long run and need to be combined or reconciled with other activities, one example being the recycling centre.

6.2.2.4 Negative issues of the learning process

One major fact that was largely stressed was the lack of relevance of EMAS within the national key performance targets. This is seen as a major burden for the future development of EMAS within local authorities in the UK. This lack of recognition is also criticised by the interviewees in regard to the European level. The use of an environmental management system should be a more important criterion within European tendering. Additionally, the interviewees demand more involvement and activity by regional stakeholders because the environmental statement is seen as a good instrument for information, though there is not enough feedback and discussion about the organisation's goals and achievements with EMAS. Thus, the idea of leading by example is not fulfilled, not because of a lack of activity, but because of a lack of communication and interest.

Regarding the local political scene, some interviewees expressed that if the political majority were to change this would obviously have an effect on the environmental policy's importance and would be a danger to EMAS.

(Voice025.done-110): "So, if we had problems financially, then the first things we would have to consider are the non-statutory functions, of which EMAS is a part. And I think if our political make-up changed, then that could be reconsidered anyway because there may not be such a vital view placed on the environment."

The EMAS regulation is criticised because it does not relate to the specific situation of public authorities. Here, the interviewees mentioned their wish for an improvement through the upcoming changes of the regulation.

(Voice025.done-168): "[…] I think one thing that would be of improvement we think is the regulation to have an element that has better inclusion of councils. 'Cause it's very much geared towards businesses in EMAS. And I think it does not really reflect the situation of the local authority, as well."

Several intra-organisational problems of the EMAS implementation and the learning process are stressed by this council as well. One is the fact that changes of the internal structure of the council sometimes make work difficult because all parts of the council except the senior management team have to work as profit centres. Another is the political pressure to improve the cost-benefit ratio of EMAS although the interviewees admit that no real cost-benefit analysis has been done so far. A third problem is staff changes that lead to a discontinuity of several projects and policies within the council.

6.2.2.5 Overall conclusion of the implementation process within this council

The council is a good example of an authority where the political will paved the way for a success story regarding the use of the European environmental management standard EMAS. The long-lasting majority of the Liberal Democrats led to a stability of the environmental policy over more than a decade. The continuous commitment to invest in environmental activities in order to do more than minimally necessary made the relatively small organisation one of the leading ones in this policy field. The analysis made clear that it is a combination of political commitment, a positive attitude towards environmental issues, the use of necessary programmes as well as a professional organisation and support structure that is rooted within the organisation which have led to a positive result regarding the implementation processes of EMAS.

Despite that, EMAS has its problems within the authority. The environmental commitment is under constant pressure from the financial side and lacks support from politicians, business and industry and society. While the policy instrument EMAS is embedded in the internal strategy of the council, it lacks the inclusion in other policy fields or instruments. Therefore, the future of the policy programme is unsure in the long term. Consequently, on the one hand, the analysis has made it clear how far the council has come and what is necessary to achieve this high level of environmental management but on the other hand, the limits of the policy programme are clearly visible as well.

6.2.3 Authority UK 3

6.2.3.1 Concept of public authority

Like the other two UK authorities in this study, the council is modern and efficient and works using the 'best value performance principle', delivering services to the people according to their needs. The council is outcome orientated; its activity guiding principle is adaptation, the organisation works according to final programming, transparency, efficiency and the national key performance indicators. Problems are solved by the termination of business and/or a legal process. Regarding EMAS, officers are partly intrinsically motivated because of the importance of the environment for tourism, which is an important economic factor in the region. Furthermore, the high environmental awareness of local residents contributes to the eco-activities of the authority. The organisation has a flat hierarchy and regarding EMAS, the council's working group is given relative freedom to work because environmental issues have a high priority.

6.2.3.2 Policy cycle of EMAS introduction and implementation

Within the *problem perception phase*, the actual impetus came through a presentation of the management system to senior officials of the council. This presentation was made by a quango organisation, the LGMB, largely founded by central government, which had the task of supporting local governments with their organisational management reforms within the process of modernising public authorities (see chapter two for details). Another driving factor was the growing awareness of environmental issues in the middle of the 1990ies. The interviewees did not express a discrepancy between their perception and other people's perception that led to the introduction of the EMS.

The *agenda setting process* can be described as partly adoptive; there was neither much support nor other organisations to give orientation on how to work with EMAS. In many ways, it was also an innovation for the organisation to introduce the management standard because the EU regulation only provides the framework which every organisation has to fill with life.

The *policy formulation* was done by the later EMAS coordinator together with the council's environmental committee with the support and supervision of the senior management team. Within the senior management team, the strategic director has profound knowledge of EMAS because he used to work for the organisation that promoted it in the 1990ies. Therefore, since this person took up the post within the organisation, the management system has been represented and supported among the senior management team of the organisation.

The *decision* to implement the system was made by the council as well as the chief executive, while preliminary decisions were made by the EMAS coordinator together with the senior management and the environmental committee. Due to staff changes throughout the whole time EMAS has been used, there are no clear indicators either for the closed nature or plurality of the decision-making process. Further, the interviews did not give any information on this point.

The *policy implementation* was led by the EMAS officer. When this post was abandoned after the introduction period in 2001, some tasks were transferred to a policy officer who has a broader job description and works on policies and strategies but closely cooperates with the strategic director. Some work is done by a consultant who is hired whenever needed, usually for ten to fifteen days a year, especially when audits and/or the environmental declaration have to be prepared. The policy officer works with the policy instrument; the chief executive is responsible for the allocation of the resources. As a special case, the authority has close contacts with some councils that either have EMAS already or were/are interested in using it.

For this authority, EMAS is closely connected to its waste collection activities because this is one of the major issues in its environmental policy, resulting in one of the best recycling rates in England. The chief of environmental affairs works together with the environmental committee on the policy programme's concepts while the (part-time) policy officer is responsible for the implementation. The other officers work with EMAS through their daily work. All staff members as well as the general public are the addressees for the programme. The influence on other programmes is especially visible within the field of energy procurement, where the council has invested in an energy company that provides electricity produced with a wind turbine for the organisation's use. Other fields are procurement, where the council takes ecological impacts of goods and services that are procured into account, and waste management, which is closely linked to the council's efforts to deliver a more ecological waste collection by introducing collections for glass, paper and plastic materials.

The ex ante and ex post *policy evaluation* is part of the EMAS management system, as is the environmental statement. This policy evaluation is carried out regularly according to the EMAS guidelines. The political learning of this organisation is the implementation of EMAS into the overall management system as well as the combination of the different policies, programmes and schemes – nationally or regionally – to form a useful combination. The latest environmental statement is a good example of this. Here, the connection between the EMAS management system, the national performance indicators and other policies is clearly visible.

Up to the time of the interviews, the organisation had no intention to *terminate* the work with EMAS; the senior manager expressed the commitment to continue. In the first quarter of 2007, the organisation published its environmental statement for 2005/06. The next revali-dation was due in 2007. Up to the fourth quarter of 2008, there was no detailed information available on the current status of EMAS within this organisation. It is still registered within the UK EMAS register and communicates its EMAS activities on the website but has not recently published an environmental declaration. No other information is available from the organisation.

6.2.3.3 Organisational learning

The espoused theories provide insight into an environmentally aware organisation that is active in many fields and has close contacts with councils that have EMAS and those that are interested in it. The theories in use reveal that the organisation is environmentally active and is working hard to meet all the necessary requirements. Regarding other councils that have EMAS, the organisation claims to have contacts with these while during the interviews, it became clear that the council does not even have an overview of EMAS' situation within the UK.

Within the *single loop learning*, the outcomes are technical improvements regarding waste collection, energy consumption and the isolation of council housing as well as a changed procedure for monitoring and billing waste, water and energy. In comparison to the other organisations of the study, the council is very active in the fields of waste minimisation. Examples of this are the programme to deal with abandoned vehicles within a very short time and the policy for environmental waste management concerning building waste from the council's own premises.

Within the *double loop learning process*, the following outcomes have been collected: the organisation has created a database for environmental management and developed a better communication with the county council about the procedures and the management of waste disposal issues. There are no direct or indirect statements of the interviewees concerning the work with the regulation itself. Nevertheless, over the years, the council has changed its management structure from a full-time EMAS coordinator to a part-time policy officer and additional support from a consultancy whenever needed. In line with this development, there is a tendency to reduce the documentation needs to a minimum. These are reactions to the experiences made with EMAS over a long period of time. As part of other aspects of learning with EMAS, one can see that the organisation has learned to reduce their resource consumption in the main building. These outcomes are now being applied to other sites of the organisation.

On the *deuteron learning level,* the interviewees expressed their belief that they could have done all the things they did without the help of the policy instrument EMAS but without EMAS, they wouldn't have done anything. The system gives the needed framework and puts pressure on the actors to become active. EMAS was the organisation's first management system. Over the years, it was integrated into an overall management system for the council. The environmental issues are only one set of issues among numerous other management issues. Regarding the EMAS management system itself, the organisation has learned how to use it and to minimise its efforts to continue with it. As part of an EMAS follow-up project, the energy strategy of the council was reconsidered and it was decided to invest in a wind energy company that would produce the whole energy of the council solely through wind power. A second project was the distribution of a recycling box to some parts of the council's area. The response of the general public was overwhelming and, consequently, the council introduced the recycling scheme to all the households in the region.

6.2.3.4 Negative aspects of the learning process

The interviewees remarked that some of the suggestions made by the verifier were not seen as very helpful. Therefore, they requested a more detailed look at the specific issues of a public authority. Furthermore, the lack of integration of EMAS in the national performance indicators was an issue. The organisation has therefore tried to overcome this on its own, at least within their internal management system.

6.2.3.5 Overall conclusion of the implementation process within this council

The council is a positive example of a public authority that has taken on the voluntary task of working according to an environmental management system.

The analysis clarified that although there have been some problems, the management system has been a success for the organisation, its staff and, of course, the environment. For this council, EMAS is a policy instrument that provides a structure for the political will of the council and the enthusiasm and activity of the staff. This is, according to a senior officer, the core of the system and the main reason for having it:

(Voice003UKok-34): "I think politically, the council has a priority around environment; politically, they are able to point to the year-on-year improvement in the council's environmental performance. So politically, it works because it provides a mechanism for saying this is important, this is how we are putting resources in, and this is some of the outcome we get in from that process. And it works managerially because it continues to keep, if you like, keep people focused on the political priority around environment but making sure they are continually trying to improve what they are doing environmentally."

Nevertheless, as with the other authorities, the lack of integration of EMAS into the national environmental policy strategy and the negligence of the performance improvements that are achieved through an environmental management system reduce the policy effectiveness of the EU regulation. This is a long-term, strategic disadvantage for EMAS and, thus, also affects the overall view of European environmental policy.

6.3 Authorities in Germany

6.3.1 Authority GE 1

6.3.1.1 Concept of public authority

The organisation's aim is to be modern and efficient. Its structural guiding principle is legality; it is orientated towards input, partly towards outcome. The activity guiding principle is stability, while the type of programming is conditional. The organisation's modes of work are transparency and documentation. Problems are usually solved through a legal process. The officers are partly intrinsically motivated. Regarding environmental management especially, the EMAS officer and members of the working group express this motivation. The organisation has a hierarchical unitary structure, although the EMAS working group has a policy programme focus and, thus, some room to work on its own.

6.3.1.2 Policy cycle of EMAS introduction and implementation

The *policy perception phase* started when the later EMAS officer felt that the work in the field of waste management and the other environmental activities of the organisation – internally and externally – needed a coherent structure and a common approach. This person learned about EMAS at a conference organised by an authority that already worked with the management system. The information gathered at the conference was discussed with the senior management of the organisation who, in the end, was convinced of the use of EMAS and started the political process to implement it.

The *agenda setting process* can be described as adoptive. Compared to the UK, there were more authorities in Germany that were working on the implementation of EMAS at the time when this authority started its environmental management project. Furthermore, there were several brochures, studies and books by regional and national authorities as well as research institutions available to get detailed information (see chapter three for details). Nevertheless, there was also the element of innovation within this process because the EMAS management system had to be customised according to the specific needs of the organisation.

The *formulation* of the environmental policy as well as the environmental programme was done by the EMAS coordinator in cooperation with the senior management team. These actors were supported by an experienced intern who had studied environmental management at university and worked on this project phase for three months. Within this phase, the actors used information issued by the regional environmental agency as well as from other councils that were already working with EMAS. The coordinator had a personal interest in the project because it promoted and secured the officer's position within the organisation. Furthermore, the role of this member of staff changed from an officer within the traditional line structure to one with close contacts to the senior management team.

The *formal decision* was made by the council of the organisation, while preliminary decisions were made by the chief executive together with the responsible senior manager. The preliminary process was regarded as one of plurality whereas the final decision was a closed one because the chief executive decided to implement EMAS within the authority.

The *policy implementation* was led by the EMAS officer under the responsibility of the chief of the environmental department. The EMAS officer works with the policy programme; he is supported by an inter-departmental EMAS working group. The allocation of resources is determined by the chief executive but has to be agreed within the financial planning by the council. Over the years of work with EMAS, the organisation has been active in the promotion of the system through participation in regional and national conferences, workshops and publications as well as numerous other promotional activities. Further, the organisation led a regional project to introduce EMAS at other authorities.

The environmental programme and the environmental policy were developed by the responsible senior manager together with the EMAS officer, who was and still is also in charge of the implementation activities. The other officers come into contact with the system through their own work or when working together in the EMAS working group. All staff members, the general public as well as affiliated organisations of the Landkreis, particularly if they take over parts of the management system, are the addressees of the programme. The implementation of the environmental management system has had an effect especially on the procurement and the waste management policies of the authority. Procurement is done under environmental aspects whenever (organisationally and financially) possible. The interviewees agreed that the recognition of these aspects is often very difficult because information to evaluate the environmental aspects and effects of the relevant product or service is often not available or very difficult to estimate.

Through the need for ex post as well as ex ante *evaluation* as well as a regular reporting, the environmental management is one of the best evaluated and documented policy programmes of the authority. This is generally true for the German authorities, especially compared to UK authorities that have to work according to the national performance indicators. For the German local authorities, there are not even indicators set by each Land. The interviewees agree that the regular validation through EMAS puts on a pressure that keeps the system moving:

(Voice016-133): "Because there is pressure within EMAS due to the validator, there is a need to do something. If that pressure were not there, not that much would have happened."

The intensive and regular evaluations lead to a better communication of staff in different units on how to work with and improve the system due to its cross-departmental and cross-thematic approach.

At the time of the interviews in April 2005, the organisation committed itself to *continue* with EMAS in order to improve the system and reduce their environmental outcome. Since that time, it has continued to do so. The latest validation was done in 2007; the next one is due in 2010.

6.3.1.3 Organisational learning

The espoused theories of this organisation imply an environmentally aware organisation that will get a better image through the use of an EMS. However, the theories in use show that although the authority is environmentally active – even more active than others, there is not much feedback of external actors on EMAS. Furthermore, internal problems between different departments lead to frustration because the environmentally eager staff cannot work as they are supposed to do due to different policies that are not harmonised. One example is the effort to reduce the energy consumption of the IT systems. While the energy savings officer wants the systems to be totally disconnected from electricity every night, the IT department demands a permanent stand-by modus in order to be able to access the systems at any time for servicing works.

The *single loop learning* reveals that here – as well as within the other organisations of the study – especially technical improvements to reduce the environmentally harmful outcome of the organisation in the fields of waste, water and energy use have been done. Indirect environmental aspects are not in the focus and are tackled only to a minor degree.

The *double loop learning* processes can be ordered among the following four main topics that have already been used at the other organisations. First, the collection, analysis and administration of all relevant data for EMAS have been organised. This data warehousing is a necessary tool for the successful implementation and continuity of EMAS and was completely new to the organisation. Second, the communication structures have been improved, especially among the units that had and still have large environmental impacts or are very active with the management system. Third, the target setting and time-sensitive reporting as tools to manage processes were new to this policy field and have improved the communication and the flow of management processes. Fourth, now that these management capabilities are available, the council members demand more detailed information within a shorter period of time.

Over the years, the organisation has developed a great deal of knowledge on how to work with and interpret the relevant EU regulation and its recommendations as well as other guidance documents in connection with EMAS. Whereas at the beginning, the regulation was followed by the word and led to a large amount of work and large files being produced, this has changed over the years. Consequently, the quantity of time and work needed for documentation of the management system as a whole has decreased while the quality has increased. Thus, the work with the regulation is a good example for the learning process that is necessary when using the management standard. The learning outcomes can and are being used on other policy instruments. Nevertheless, the overall policy learning capability could at large be improved and thus would lead to a more flexible and more successful strategy in environmental issues and beyond.

Another effect of the management system was the reduction of environmentally relevant costs. About 50.000 Euro were saved in the period of 1999 to 2004 through the work with the management system, excluding internal staff costs. While in the first years of EMAS, cost reductions were relatively easy to achieve, over the years, the savings per annum to be achieved have generally decreased while demands for investments to achieve more savings have increased because, over time, the more severe environmental impacts have become visible.

Regarding staff involvement alongside the EMAS process, all staff were informed, encouraged to take part in the process and got regular internal training on environmental issues. Some staff members even had the chance to get an extra qualification for this new task – especially the EMAS coordinator. These new qualifications led some staff to be more independent and more actively involved in their daily tasks. The continuous involvement of staff is a key issue for the management system.

Among the *deuteron learning* processes, the first item is the overall learning of the fact that the work with EMAS improved the cross-departmental communication as well as creating a greater environmental awareness that is the basis for a new perspective on the individual's work. Some interviewees even expressed that this new environmental awareness had an effect on their private lives and/or staff members' private lives and behaviour. Regarding the management system, the organisation realised that the environmental activities needed an overall umbrella that would bring the different activities and policies together. After almost eight years of work with EMAS, the conclusion is that the EMS has delivered this overall function.

The authority did not only use its capabilities of environmental management for its main site, where EMAS was implemented, but took part in projects to pass it on, both within its organisation and also beyond. It took the lead in a project to implement EMAS in other local authorities, implemented parts of the management system at other sites and buildings and passed information on how to work with EMAS on to other interested organisations.

(Voice17-116): "This model project was planned to take place in all districts of Bavaria. Its aim was and is to distribute EMAS throughout the whole Land. In the end, only authorities of Southern and Northern Bavaria participated in the project. We were the lead partner with four authorities that wanted to introduce EMAS."

In addition to the developments with EMAS described so far, there are external effects of the system as well. A central one is that companies try to advertise their environmental certificates, usually not an EMS but lower-level systems, to the authority because they know that the Landkreis is environmentally active. Through showing their environmental activities, they hope it will be easier to develop a business relationship with the Landkreis and get more contracts.

6.3.1.4 Negative issues of the learning process

According to the present governmental accounting system of the council, a cost-benefit-analysis of EMAS is not possible or necessary and has, thus, not been done so far. Although, if done, the financial input needed for the system would surely outweigh the financial savings achieved over the last years. This is known to the organisational management. Only the external costs were presented to the council during the last years. Thus, the impression was that the authority gained savings through the environmental management system. In regard to the NPM approach, especially compared to UK authorities, this lack of financial awareness can be seen as highly critical. In addition, the EMAS coordinator claims that over the years, it has sometimes become difficult to motivate staff to continue with the system because recognition and awareness of environmental issues have decreased. In addition, processes of the development of new procedures or the implementation of these often take longer than expected. The interviewees feel that sometimes these processes take too long and therefore reduce the current motivation and the activities of the members of this authority.

Another problem that also had been mentioned by other authorities was the collection of data that was needed for the management system. The amount of data needed meant that the authority sometimes had difficulties collecting and analysing it because of the minimal staff available.

The environmentally-orientated procurement is an issue where improvements would be useful. Although the authority is working according to environmental standards for its procurement, there is no contact and exchange of information with other authorities who might work in the same direction or have additional knowledge that could be helpful. There is a lack of cooperation and coordination regarding problems and solutions that would be of interest for a large number of authorities. Here, a regional contact or coordination office or even a more centrally-organised service would be a good way to promote ecologically-orientated procurement.

This development is in line with the decrease in contacts to other authorities on EMAS over time, while within the first years of EMAS, responses to the environmental declaration and requests for assistance or information on EMAS were more frequent. In total, the EMAS coordinator had most requests and responses from the scientific community and the least from the general public. These reactions resemble an overall trend that EMAS is not of great interest, in part because it is not well known but more because it is an internal system having no major effects outside the organisation. Again, the lack of interest is an indicator for the low profile EMAS has with the general public. What is even more important is the lack of policy integration of this NEPI tool with other policy instruments.

6.3.1.5 Overall conclusion of the implementation process within this council

Although the council's concept of public authority can be described as classical Weberian, the analysis has revealed that within the policy field of environment, the council has been working with an NPM-orientated management instrument for eight years now. EMAS was the organisation's first management system. The council has been successful with the system, although there are some problems with it. Nevertheless, there is rich experience with the management system which has been transferred throughout the organisation and beyond. EMAS has become part of the organisational culture now and has influenced some parts of the organisation, thus changing the behaviour of the organisation.

As a result of the analysis, the system is facing three challenges for the future: First, it needs more response and recognition, both from within the organisation and beyond. Second, the organisation has to decide whether or not to tackle indirect environmental aspects. Third, the management system needs to be more integrated into an overall modern management system covering a whole range of issues. Fourth and most important is the aspect of policy integration that has to be strengthened within the organisation and in the environmental policy of the Land as well as of federal government. These aspects together will bring the system and its achievements forward for the organisation. If the authority does not consider developing the system any further, the question of its usefulness will shortly come up, especially if the financial constraints of the organisation continue. Therefore, a development plan for EMAS is essential for its future within the organisation.

6.3.2 Authority GE 2

6.3.2.1 Concept of public authority

Like the other two German Landkreis authorities within this study, the authority's aim is to be modern and efficient. Its structural guiding principle is legality and the orientation is both towards input and outcome. The organisation's activity guiding principle is stability, not adaptation, like at the English counterparts. The type of programming is conditional, the mode of work and the documentation follow the principles of transparency. Problems are solved through a legal process. The attitude of the officers is partly intrinsic; regarding EMAS especially the main actors show an intrinsic motivation. The organisation's structure is hierarchical and unitary although the EMAS working group does not resemble these structural principles; it is cross-departmental, its members are chosen in view of their capabilities, not because of their status or role.

6.3.2.2 Policy cycle of EMAS introduction and implementation

The *policy perception phase* can be described as a search for a coherent and overall structure for all the environmental activities. This generated out of an interest developed by the unit responsible for waste management and environment because staff here felt that there were different policies that were not managed according to a coherent plan. Furthermore, the organisation wanted to improve its own environmental impact to show its capabilities and represent a leading role within this policy field.

The *process of the agenda setting* can be described as mainly adoptive because when the EMAS project began, there were already several local authorities in Germany of different size working on it and information about the implementation of it was available. Nevertheless, as with the other authorities, one can also describe the process as innovative because the organisation had to invent new procedures according to the EMAS regulation for its own purposes.

The *policy formulation phase* was mainly the formulation of the environmental policy and the environmental programme according to the EMAS requirements. This was done by the later EMAS coordinator, an environmental engineer, in cooperation with the senior management team. This person had a personal interest in the project because it brought together several aspects of the environmental activities of the organisation as well as helping to promote their status within the organisation.

The *decision making process* was lead by the council and the chief executive. The latter, together with the senior management team, was responsible for preliminary decisions of which the first, regarding the introduction of the management system, was taken in a closed process because it was taken among a very limited number of members of the senior management. In contrast, the ongoing work with EMAS is considered as pluralistic, especially among the members of the working group. This group represents the EMAS aim of being cross-departmental and the idea of working with it across policy fields.

Within the *implementation phase*, the EMAS officer was and still is the focal point of activity. Together with the EMAS working group, he has worked on the policy programme on a continual basis over the years, i.e. the management of the programme, the preparation of the validation periods, information of all relevant stakeholders, whereas the chief executive is responsible for the allocation of resources. The cooperation and coordination of the team members coming from different departments is a key to a successful EMAS implementation because these officers usually have the knowledge and motivation to develop activities for the organisation as a whole as well as for single departments. The EMAS process within this authority is closely linked to the organisation's waste management activity because the first impetus for the need for an EMS came from the waste management department.

Regarding the roles of the different actors according to the policy cycle theory, the EMAS officer is the programme conceptionalist. At the same time, he is also the implementations manager. The other officers working with the management system within their daily routines are the street bureaucrats, while the addressees of the policy programme are staff, general public and staff of organisations affiliated to the Landkreis (i.e. also in the main building or those who have a close cooperation with the authority). The implementation of EMAS has had an effect on the following policy programmes: procurement, energy management and waste management. These policies have been integrated into the management system or have undergone considerable change regarding the work on the relevant policy problems alongside the environmental management.

The *policy evaluation* is part of the PDCA-cycle concept of EMAS that requires both ex post and ex ante evaluation. This is done regularly. Therefore, as with the other authorities, the environmental management system is one of the best evaluated policy instruments. Although the systematic evaluation of the policy instruments and its outcomes is generally welcomed by the central actors of EMAS within this authority, the interviewees remark that it is a complex and long-lasting task. They would prefer less strict rules but agree that through these, the effectiveness of the system would be reduced. Regarding the political learning, the system was the first management-orientated instrument of the organisation. Over the years, it has lead to the idea that a general quality management system would be helpful to have an overview of the standards of quality the organisation is delivering. At the time of the interviews (in April and June 2005), the council was considering the introduction of quality management according to the ISO 9000 standard and the integration of EMAS into this quality assurance system.

The *policy continuation* is ongoing. The organisation has a strong commitment to the system. Environmental issues are very important for the organisation. This is visible not only through the work of EMAS but also through a number of other activities such as energy consulting services for other authorities like town councils as well as the general public, a large support for the Local Agenda 21 process as well as activities providing environmental education for schools and pre-school organisations. There is no intention to terminate EMAS. The latest revalidation that was due was achieved in May 2007.

6.3.2.3 Organisational learning

The espoused theory of this organisation reveals a modern, environmentally aware organisation that intends to improve its accountability and image through the EMAS system. Additionally, the system has made staff proud of their achievements, especially through the public accountability effect.

The theories in use give the impression that although the organisation is environmentally active, it is not sure how well and widely EMAS is recognised because there are very few reactions from the general public on the environmental activities. When the council tried to interest local and regional companies in the idea of environmental management, most responses were negative. Consequently, a planned project to share the knowledge about EMAS with others in order to implement it in companies and other organisations within the region did not take place. The main problem is that although the organisation is environmentally active, these activities are not perceived as being important for the general public.

Within *the single loop learning*, technical changes i.e. change of light bulbs, the use of energy saving pumps, etc. to save energy, changes to save water and waste and the collection of waste paper were made through small scale technical improvements or the change of procedures and routines. Remarkably, these activities are visible within the building through numerous small signs and information boards.

The *double loop learning* analysis again revealed four types of learning. One is the data collection and the creation of a database. Regular data collection, analysis and monitoring helped the organisation to reduce costs and react to changes and unforeseen circumstances quickly.

(Voice033-77): "A positive thing of EMAS is that one has to publish an environmental record once a year. It uncovers new and unexpected things every time that otherwise would not have been noticed. For example, with energy consumption. If one notices an increase in electricity consumption over the last months, one can take a close look at the causes of this increase. Without the environmental record, nobody would have noticed this."

This data collection was the basis for an improved communication on the environmental impacts of the authority and is also partly responsible for an improved internal communication.

Through the work with the regulation, the authority has learned how to understand and interpret it, i.e. what is important, how something has to be done, what is compulsory and what is voluntary. During the time working with EMAS, the authority has also learned to decrease the amount of documentation and formal rules due to a better understanding of the regulation. The pressure through the external verification process of the policy was described as helpful by the interviewees because it was encouraging for the whole EMAS team and led to a quickening of the implementation process. Other effects of the work with the EMS were that the members of the organisation were more cautious about their energy consumption. In general, a greater knowledge and awareness about the environmental outcomes of the council's activities was noted.

Regarding the overall learning, the *deuteron-learning activities* revealed that with EMAS, a new culture of coping with management systems was introduced in the authority. The learning outcomes of the EMAS processes led to the introduction of elements of a quality management system, although the authority does not aim to achieve a certification:

(Voice033-27): "Of course EMAS has large advantages, but it also has its disadvantages, for example the bureaucratic needs like the obligation to document everything, etc. These are sometimes quite clamming and cause a lot of work that would not be necessary. Sometimes one could save a lot of work and energy. And with the quality management, we now go another way. We choose methods and instruments that we think are relevant but leave out the verification and other duties that would be necessary with a standardised quality management system."

As a central follow-up project, the authority initiated the introduction of low-level environmental management systems for companies in its region. This was also due to the negative response to EMAS by these companies because they were of the opinion that it would be too complex.

As a side effect of EMAS, an improvement system was introduced, leading to about fifty improvements of operation within the authority over the years. The management system has led to a culture of more control, a better communication among the staff members about environmental issues and the aim to improve the outcome of the authority. These elements have spread from the environmental policy onto other policies of the organisation (where it has the legal capacity to decide on itself).

6.3.2.4 Negative issues of the learning process

First, there is not much knowledge about EMAS in other organisations. Furthermore, there were few responses from the general public when the authority made its commitment and achievements with the system visible. Second, the management system itself is seen as quite bureaucratic; over the years there was a need to reduce the amount of documentation and paper-work to make the system manageable. In addition, the internal and external validation is seen as quite a burden for the authority because it has to be done on a yearly basis:

(Voice033-79): "Twice a year, we are obliged to do an internal validation of two departments. In addition, there is the external validation. As a consequence, one is constantly working on EMAS and it is an ongoing review process."

Third, the experience with EMAS revealed that the auditor was asking for activities within the framework of EMAS that the organisation was not willing to do. The cooperation did not seem to be fruitful. Therefore, the organisation changed its auditor. With the new one, the following validations went well because this person demanded a working management system and waived several activities that the former auditor had demanded for a successful validation. This is a sign of a different understanding of the EMAS regulation by auditors which can be seen as critical, not only in the national but even more in the international perspective.

For some members of staff, especially the lower technical officers, the management handbook seems to be too complicated and is, therefore, not used very often. Thus a need for a less complex document is raised. Further, the interviewees explained that solution finding sometimes takes longer than expected during the work on a problem, a fact predominant in environmental issues. This is due to the complexity of some issue that becomes visible during the problem solving process; sometimes alongside a lengthy decision process within the authority. Some interviewees even expressed the lack of a personal reward if one takes an active and/or responsible role in the management system. This leads to a decrease of interest in the system because the environmental management system is seen as an additional activity on top of the officer's daily work without any given time resources. Further, the lessons learned from the management system are not systematically transferred to other sites of the authority. Nevertheless, the authority is financially supporting a regional energy and resource reduction centre. This centre will do some in-depth research on the energy consumption in the future (at the time of the interviews). All these factors lead to a decreased interest and willingness to work with EMAS, which is seen as extra work by a large proportion of the authority's staff.

6.3.2.5 Overall conclusion of the implementation process within this council

The authority has been working with EMAS for nine years now. It has successfully implemented a management system that was new for the organisation. Over the years, it has integrated it into its other policy programmes – not only the ones from the environmental policy field. With this system, it has achieved a higher level of structured environmental policy as well as a higher reflection of its environmental impacts than without it. There are several reasons for this. One is the support by the chief executive and the senior management team which has continuously been there since the beginning. Another is the professional management structure as well as the education of the environmental manager. Being an environmental engineer in the waste management department, this officer has a central function within the system. His intrinsic motivation is reflected in his work and his commitment to the EMAS system. The long-lasting political stability of the council's senior management team as well as the council's majority is also a factor contributing to the long-term commitment to EMAS.

Nevertheless, the very few reactions to the authority's environmental activities and its use of EMAS have led to a difficult situation of the policy instrument because the image that the organisation wanted to produce as well as the recognition by the public that were projected by those organisations that promoted EMAS did not materialise. Thus, these would be important external drivers for the continuity of the programme because positive public recognition of a policy programme directly affects the ideas of the relevant actors who have to decide on the future of EMAS. The lack of personal rewards, the failure to integrate the system at other sites of the organisation as well as the lack of a cost-benefit analysis is burdensome for the future of the system.

The policy instrument EMAS within this authority is an internal policy programme for the improvement of the environmental performance and the development of management procedures. In its use within this authority, it cannot be described as a classical instrument of NPM because there is no budgetary control concerning EMAS. Thus, one central issue of the NPM instruments, the cost-effectiveness, is missing. Nevertheless, EMAS has been successful at this organisation because of the general setting of the policy programme, although there is potential to improve and expand the programme.

6.3.3 Authority GE 3

6.3.3.1 Concept of public authority

Unlike the other two authorities, this one is organised according to the classical KGSt-structure that was proposed to local authorities until the early 1990ies. This concept of public authority represents the one of the 1980ies and 1990ies in Germany, before the discussion of NPM. This is due to the special situation of the public authorities in the Eastern German Länder that had to be restructured in the early 1990ies as a reaction to the re-unification process and the changes of administration that followed. These administrations are facing a new round of structural reforms from 2007 onwards due to the economic and demographic problems of the region. The authority's aim is the classical Weberian orientation of public authority; it is working according to the principle of legality. The orientation is focused on input, the activity principle is stability. The type of programming is conditional; the mode of work is transparency and documentation according to the rules. Problems are solved through a legal process. The officers' attitude – especially towards EMAS – is impersonal and orientated towards routines. The organisation's structure can be described as hierarchical and unitary. Regarding EMAS, there is no permanent working group.

6.3.3.2 Policy cycle of EMAS introduction and implementation

The *problem perception* began with the idea that the organisation wanted to do its own environmental project in the course of its Local Agenda 21 activities rather than primarily acting as an organiser for projects. Through the help of an external consultancy that is owned by the regional public authorities, the organisation got state funds to start the project:

(VOICE032-2): "We came to this topic through our Local Agenda 21 activities. We have been working on projects within this context since 1998. In the course of time, we were looking for a project that we could do within the administration rather than having the sole role of organising projects. Through contacts with local companies, we got information about EMAS. In 1999, we then contacted the relevant state administration and asked for financial aid. Through the state's environmental foundation, we got some money for the project. Since 2000, we have been actively working with EMAS."

The *agenda setting process* can be described as a policy adoption, although the authority had no experience with management systems at all. The introduction and implementation of EMAS was the first time for the authority to actively implement a voluntary European legislation. There were no public authorities in the region to exchange experiences and to discuss problems with. Therefore, many things had to be done with the help of the consultancy. The environmental policy as well as the environmental programme was formulated by a consultant in cooperation with some members of staff. It was also the consultancy who helped to get funding for the project from the state authorities. Without this funding, the process would not have been possible.

The *formal decision* was made by the council together with the chief executive; preliminary decisions were made by the senior management team under the supervision of the chief executive. The decision process can be described as a closed one, being a top-down decision for the organisation.

The *policy implementation* was supervised by the chief of the economic affairs department. The main work was done by a consultant together with one part-time office worker and an ad-hoc EMAS working group. The chief executive was responsible for the allocation of resources. In contrast to the other organisations of this study, the authority had considerable support from the consultancy. Even the implementation was largely organised by this consultancy.

Regarding the conceptional work of EMAS, this was done by the consultancy, as was a large part of the implementation. The organisation's EMAS officer had only a minor role in this process. Other officers together with the consultants were the street bureaucrats who had to work with EMAS within their routine work. The role of actors in this organisation differs considerably from all the other organisations in the study regarding the programme conceptionalists, the implementations managers and the street bureaucrats because these roles were mainly taken over by the consultancy working for the authority. Only the addressees of the programme, i.e. all staff members of the organisation as well as the general public, are the same as with all other authorities. These differences in the role model have given way to a rather superficial work with EMAS; it is a primary top-down process, done by external actors, leading to a rather formal compliance of the environmental management system. EMAS had a major effect on the premises of the organisation because they were renewed due to the work with the environmental management system. The renewal process would have been necessary anyway but it was easier with EMAS because the authority was able to get additional funds for this project.

The *policy evaluation phase* of EMAS requires the ex post and ex ante evaluation. This was mainly done by the consultancy within this organisation. Thus, the policy instrument and the relevant documentation seem static and not filled with life. In contrast to all other authorities, this organisation had the advantage of strong support but the disadvantage of not reflecting upon the use of the management system. Consequently, the environmental management system stayed on a rather abstract level and was not developed further or integrated into other policy instruments.

Within the interviews, no change regarding the political learning was visible. The management system was implemented in a top-down process. The responsible desk officer works in the economic affairs department, a situation that is exceptional for all authorities within this study where EMAS is usually situated in the environmental department. No systematic or thematic reason was found or given for this decision.

Regarding the *policy continuation or termination*, the status of EMAS is unclear at present (December 2008) because the authority has merged with other councils to form a new larger authority (German: Landkreis) due to the demographic and economic development of the region. Therefore, no information is available on the current status of the policy programme. This uncertainty regarding the continuation of EMAS had already been mentioned during the interviews in 2005. The last environmental declaration was issued in September 2006; presumably the project ended with the merger of the authorities in June 2007

6.3.3.3 Concept of organisational learning

The espoused theories give the impression that the organisation is partly environmentally aware and combines the Local Agenda activities within the region with the internal policy instrument EMAS. The theories in use reveal that after the initial implementation phase took place and the financial support from the Land had ended, the main idea was to keep the system running formally, pretending not to lose it. Nevertheless, the interviewees gave the impression of uncertainty of the depth of implementation as well as of the future of the system. Consequently, there is deep concern regarding the intensity with which the organisation worked with the system.

Within the *single loop learning*, technical improvements were made and procedures changed to reduce the consumption of energy and water, and to reduce waste. These improvements are typical for authorities using EMAS as we have seen with the other organisations analysed.

Among the *double loop learning* issues only very few changes can be noted. From the interviews, there is no real evidence that the council has made much effort to create a detailed database that would help to work with EMAS as the other authorities have done. The environmental declaration does not give any further helpful information on this question; it remains rather formal and descriptive. Further, there is no clear evidence that the communication among the units and the staff has improved through EMAS. Additionally, no information on the work and the development of their own way of interpreting the European regulation is visible because most of the work was done by a consultancy.

Within the *deuteron learning phase*, EMAS did have effects on the organisation. Compared to other organisations, these learning processes only focus on the follow-up projects; there were no indications of an overall learning process or a general learning of the work with the management system. Nevertheless, some follow-ups have been initiated. One is the redevelopment of the builder's yard that was heavily contaminated and needed a complete reconstruction, both of the infrastructure and of the management procedures. A second project was the establishment of a central facility management that is responsible for all of the authority's sites. Further, a central procurement office was installed as an outcome of EMAS because it became clear through the internal and external validation that that all units of the organisation did their procurement separately before.

6.3.3.4 Negative issues of the learning process

The interviews as well as the environmental declaration reveal problems with the management system. The organisation did not work out a system of indicators. Its direct environmental outcomes are noted without relation to other data and, therefore, do not give adequate information on the outcomes in relation to the use of the building, the vehicle mileage or other relevant data that changes raw data into indicators.
Although the organisation was asked by the European Commission to take part in a project on environmental activities of local authorities, it did not do so due to financial reasons. These financial constraints are negative for the

whole management system. Since there are no further subsidies for the policy programme, the organisation does not seem to be able to develop it further. One example for this assumption is the documentation needs of EMAS which are described as too large and too complex. Nevertheless, neither any activities to reduce the documentation needs nor an integration of EMAS documents or procedures into the general operating procedures have been undertaken as in the other authorities looked at in detail within this work. Another indicator for this impression is that the interviewees express the yearly validation and update of the environmental declaration as a heavy burden. In contrast, one interviewee demanded more information exchange and training for EMAS within the public authorities. The specific needs of this kind of organisation were not adequately reflected by the current information available according to this staff member.

6.3.3.5 Overall conclusion of the implementation process within this council

The authority is a classic example of an organisation where EMAS was implemented in a top-down process. The whole policy process was dominated by the external consultancy that was responsible for most of the policy phases, leading to a formal introduction and implementation of EMAS. As a consequence, the analysis of the interviews revealed that the learning outcomes differ enormously from the other authorities which have implemented EMAS into their general management system or learned and worked out more from the system.

The organisational orientation towards the classical Weberian structure, the financial constraints, a lack of continuous and cross-departmental cooperation, an orientation towards hierarchical structures and the lack of a permanent EMAS working group, as well as the top-down communication structure have, in addition, made the situation worse for the system. Together with the very limited staff resources invested in the process, giving only one desk officer of a non-environmental department to work with EMAS, the whole process of initiation and implementation of this policy instrument process seems hollow and very formal – over the years, no positive developments are visible; over time, the environmental declaration seems a mere documentation of the status quo.

Overall, this authority is an example where the ideas of EMAS did not work due to the conceptional framework within the management system is based. In contrast to the other organisations of this study that are environmentally aware authorities with an intrinsic motivation and that are willing to establish a process for learning, improving and change, within this organisation, operational, motivational financial and structural deficits determined the outcome of the EMAS project from the beginning. The environmental management tool has proven to be too complex for the authority, which lacked the capacities and the long-term political will to be open for processes of change. The unclear status of the management system due to the merger with other authorities completes the picture. The organisation did not have the potential to use EMAS as a framework to develop a fully operational environmental management system that would meet their needs and could be the framework for their environmental activities.

6.4 Summary of the empirical findings

The analysis of the introduction and implementation of EMAS can be summarised in the following table which gives information on the time frames of EMAS, organisational orientation, the policy process and the organisational learning processes (see table 6.2).

	Authority UK 1	Authority UK 2	Authority UK 3	Authority GE 1	Authority GE 2	Authority GE 3
EMAS process since	1996	1996-1997	approximately 1996	1999	1999	1999
Time of implementation	1 year	2 - 3 years	3 years (1996 - 1999)	1.5 years	2 years	2 years
Time of registration	1997	1999, ISO in 2002	1999	2001	2001	2001
Leading party at time of decision	Liberal Democrats	Liberal Democrats	Liberal Democrats	Conservatives	Conservatives	Conservatives
Leading Party at time of Interviews (2005)	Conservatives	Liberal Democrats	Conservatives	Conservatives	Conservatives	Conservatives
Style of authority	NPM/modern	NPM/modern, very environmentally aware	NPM/modern, very environmentally aware	Classical Weberian	Classical Weberian, modern approach	Classical Weberian
Support by organisation/ Authority	LGMB (no direct influence visible)	None externally, support by Local Agenda 21 group	LGMB presented EMAS to authority in mid 1990ies.	State Environ-mental Agency (information), other authorities (infor-mation)	State Environ-mental Agency (information), other authorities (information)	Financial support by State Environmental Ministry. Staff largely supported by consultancy
EMAS coverage	Headquarters building, energy, water and gas saving programme for other sites as well	Headquarters building, recycling centre	Headquarters building	Headquarters building	Headquarters building, incl. other organisations using the building	All sites with administration, rescue service and regional road maintenance facility

Work mainly done by	Working group, EMAS coordinator (2x part-time) later EMAS coordinator (part-time)	Environmental steering group headed by EMAS coordinator (full time) supported by Local Agenda 21 working group as well as ECO monitor group	Full-time EMAS coordinator until 2001. Since then: Work done by policy unit (part-time policy officer), strategic director (senior management) and consultancy, 10 - 15 days a year	Authority's Local Agenda 21 group, EMAS officer, helped by an apprentice (part time, 6 months, at initial implementation)	Initial implementation process: EMAS coordinator (full time), part-time apprentice for 6 months, and EMAS team. After implement-tation: EMAS coordinator (environmental engineer) together with EMAS team	Initial implementation: consultancy, with help from senior staff members and working group. After implementation: officer working 20 per cent of their time for EMAS, helped by EMAS working group
Policy perception	About 1996	1996-1997	1996	1999	1999	1999
Agenda setting	About 1996	1996-1997	About 1997	1999	1999	1999
Policy Formulation	1996	1996-1997	About 1997	1999	1999-2000	2000, took about 6 months
Decision making	1996	1997	1997	1999	2000	2000
Policy implementation	Focus on direct environmental impacts. At first, indirect environmental impacts. Later,	Focus on direct environmental impacts. Work at council homes, building materials, waste	Stimulated by CPA. Direct environmental impacts worked out first, then	1999-2001, first focus on direct environmental impacts, indirect environmental	Implementation intended as bottom-up process. Focus on direct environmental	1.5 years. First focus on direct environmental impacts, indirect environmental impacts: rules

	waste.	management; indirect environmental impacts: difficult to work on, support for contractors to achieve EMAS or ISO, procurement policy for electrical devices, electricity 100 per cent from renewable resources.	followed by indirect environmental impacts: 100 per cent of electricity comes from renewable resources, catering contract in accordance with EMAS.	impacts more difficult, tackled later: procurement, cleaning, canteen, problems with landscape planning and traffic.	impacts, indirect environmental impacts difficult: procurement, travel. Question of cost-benefit ratio comes up regularly.	for procurement, discussion about further activities.
Policy evaluation	Easy to motivate staff at beginning, now it is a problem, EMAS had to be implemented into a general management system. Financial savings through EMAS of 240,000 pounds between 1997 and 2005.	First phase: get it and keep it. Now: keeping and improving it, more the long term perspective. Over the years, officers have become keener to allow staff to work on EMAS-related projects. EMS in principle not adapted to local authorities.	Over the years, EMAS has become part of the management system of the organisation. It is integrated into other procedures, thus not losing its character. Important: the pressure that EMAS puts on the organisation.	Improved communication process, improved data collection, better documentation, financial savings of 50,000 Euro (excluding staff costs) between 1999 - 2005). Learning processes, improved process management,	Improved communication, improved data collection, about 50 projects have been initiated under EMAS. System is sometimes very demanding. Not all targets could be achieved due to a number of reasons. Very little	Improved communication, improved data collection, improved processes and documentation, redevelopment of a builder's yard, energy programme for schools, introduction of integrated facility management.

			Contacts with other authorities about EMAS. These did not lead to any activity. Problem of concentrating on direct environmental impacts. No cost-benefit analysis for EMAS so far.	follow-up-projects. Few responses from public or other authorities regarding EMAS. Data collection and documentation a lot of work, cost of EMAS quite high.	response from general public or other authorities to EMAS. Problems in transferring knowledge to other sites.	Lack of financial support by any higher authority, yearly checks seen as too much work, no detailed communication with public about EMAS, problems with documentation processes.
Policy Termination /Continuation	Between 2000 and 2002 interest in EMAS declined. After 2002, EMAS was on the agenda again. Org. realised that it would lose all capabilities if system would not be continued.	Organisation is determined to keep both ISO and EMAS. Future challenge is to make EMAS more cost-effective.	Between 2000 and 2002, EMAS was left behind, partly due to staff shortage; in 2003, the system was revised. Organisation wants to continue with EMAS.	Authority is determined to keep EMAS and extend environmental checks and data analysis learned from EMAS to other sites.	Authority is determined to continue with EMAS.	Organisation wants to continue with EMAS. Due to a reform of local authorities, the future for the system beyond 2007 is unsure.
Espoused theories	Environmentally aware	Environmentally aware	Environmentally aware	Environmentally aware	Environmentally aware	Partly environmentally aware

Theories in use	Keeping the system, not spending too much on it.	Environmentally active, continuing to improve.	Environmentally active, continuing to improve.	Environmentally active, continuing to improve.	Environmentally active, continuing to improve.	Running the system in a formal way, pretending not to lose it.
Single loop learning	Technical improvements	Technical improvements, strong focus on reducing waste	Technical improvements	Technical improvements	Technical improvements	Technical improvements
Double loop learning	Learning in the fields of data collection, communication and work with the regulation.	Learning in the fields of data collection, work with the regulation, other projects.	Learning in the fields of data collection, communication, other projects.	Learning in the fields of data collection, work with the regulation, other projects.	Learning in the fields of data collection, work with the regulation, other projects.	Learning in the field of other projects.
Deuteron learning	Learning in the fields of overall learning, management system, follow-up projects and other fields of activities.	Learning in the fields of overall learning, management system, follow-up projects and other fields of activities.	Learning in the fields of overall learning, management system, follow-up projects and other fields of activities.	Learning in the fields of overall learning, management system, follow-up projects and other fields of activities.	Learning in the fields of overall learning, management system, follow-up projects and other fields of activities.	Learning in the field of follow-up projects.
Other remarks	First management system. In 2002, revised, streamlined and integrated into general management	First management system, revised and integrated into general management structures. Organisation has	First management system, revised and integrated into the general management structures.	First and only management system	First and only management system. It is seen as cheap to run and cheap to maintain if staff costs are not included.	First and only management system. Authority was asked to take part in EU project. This was turned down due

	activities. Indirect environmental impacts: very difficult, EMAS led to modernisation of office blocks.	both ISO 14001 and EMAS, took part in EU ECO budget programme and other EU-financed programmes. Costs of EMAS: about 0.8 per cent of annual budget.				to a lack of financial resources. No training courses for staff of public authorities.
Next validation	Revalidation due in 2007, status unclear.	Revalidation due in June 2007, status unclear.	Revalidation due in 2007, status unclear.	Revalidation done in 2007, next validation due in 2010.	Revalidation due in 2007, status unclear.	Revalidation due in 2007, status unclear, project seems to have been terminated with merger of authorities.

Table 6.2: Summary of the data analysis. Source: Own.

7 Final Conclusions

7.1 Introduction

A detailed analysis of the organisations is provided in chapter six. Therefore the aim of the following is to draw conclusions out of this analysis. First, the conclusions will focus on the model of administration that is the underlying structure of the organisation's operation, second, on the introduction and implementation of EMAS as a policy instrument that covers (or at least aims to cover) the majority of environmentally-important effects of the organisations and, third, the effects of this instrument on the organisational learning processes. All this is done in reference to the research questions of chapter one, which are recapitulated here:

1. Is there a structural advantage in the style and operation of the public authorities that supports the use of management-orientated instruments, in this case the introduction and implementation of EMAS?

2. How is EMAS introduced and implemented in public regional authorities in Germany and the UK? What are the main reasons for implementing it? Which policy processes lead to a successful implementation result?

3. Which learning effects for the organisation have been achieved with the management system?

In a second part, conclusions regarding the policy instrument EMAS will be drawn. These will cover the use of EMAS within public authorities, the current revision of the EMAS regulation (so-called EMAS III) as well as the future of environmental policy instruments, especially the renaissance of regulative instruments in contrast to voluntary and flexible arrangements. A third part will concentrate on further research needs regarding EMAS, and the future development and use of NEPIs.

7.2 Organisational concept of public administration and its effects on EMAS

The comparative case studies of EMAS within local public authorities in the UK and Germany have made it clear that the introduction and implementation of the policy programme is based on the underlying concept of public administration that organisations are following. These concepts determine the overall principles of the organisation including the organisation's structure, its modes of operation and management, the behaviour of staff towards operational methods as well as procedures for policy continuation and/or termination. Where the organisation is working with NPM-instruments, focussing on key indicators and a strong outcome orientation, EMAS will fit in smoothly because it is based on the same procedural ideas as other management-orientated instruments. At authorities working according to the classical Weberian model of administration, EMAS can be seen as one (or for some organisations, the first) policy programme to start new management processes which are in line with the NPM set of policy instruments for modern governance. Due to its cross-sectoral approach, numerous departments and policies can be influenced by EMAS if it is implemented on a large scale within a local public organisation, especially regarding indirect environmental effects of the organisation. The analysis of EMAS within the six public local authorities has made it evident that the environmental management system is primarily a tool for decreasing the negative environmental outcomes of an organisation. Within a long-term perspective, it is a management tool that leads to improved management capabilities of an organisation, affecting numerous fields of activity, not only the environment. Where the authority is changing its concept of operation, EMAS can be supportive if it is implemented well and is included into the existing policy programmes and communication structures.

The results of the analysis reveal that the British authorities have advantages in using EMAS because they are management orientated and work according to NPM strategies that are guiding local public authorities nationwide. At the same time, some of the instruments introduced with NPM, like the national key performance indicators, do not take EMAS into account. Consequently, this lack of policy integration has enormous effects on the acceptance of the instrument among stakeholders. Further, it affects its current use within organisations and will consequently influence the future development of the policy instrument because through the national performance standards requirements, local authorities concentrate on the centrally-defined outcomes and audit results, which are already demanding. Thus, most of them hesitate to do even more than that. If a policy instrument (or at least the outcomes achieved with it) is not recognised within the policy strategy at all, it will be largely neglected. Therefore, the use of EMAS has its origin in the intrinsically-orientated political ideas and preferences of local actors that see the system as a useful umbrella for their environmental activities. While the national influences on the use of a policy instrument are not against the European regulation's ideas and could even be in line with the European Commission's aspiration that it is a premium environmental management system, these national influences lead to a very low uptake of the system and, thus, question the overall use and effect of EMAS as a whole on the environmental policy outcome within the UK.

For the German authorities, the situation is different. The federal structure determines a rather individual selection of the public administration approach by each organisation, leading to a situation where each organisation can continue operating according to the classical Weberian model of authority, proceed towards the modern NPM approach or develop something in between. Thus, the situation of the German authorities can be described as partly in transition towards a stronger NPM orientation, depending on the preferences of the organisation's legislative and executive institutions. Although the intensity of transition differs considerably, EMAS has had its effects in Germany. First, the number of users is quite large in Germany compared to other EU member states (and, consequently, compared to the UK) and, thus, the effects of change are larger measured in numbers despite its uncommon approach for the traditional German environmental policy. All authorities of this study reveal that the introduction and implementation of EMAS largely depends on the individual preference of relevant actors in the senior and the middle management. However, the German ones are structurally not that much determined to national standards that influence their policy activities compared to their UK counterparts. In addition, for the German authorities within this study, EMAS is the first policy instrument with a management-orientated structure; therefore, it is seen as a challenge or as a burden, depending on the learning and management capabilities of the each authority. Often, EMAS provides the first stage of a modernising process within the field of environmental policy whereas the UK organisations have an advantage through their clear focus on NPM instruments that also covers environmental issues. Furthermore, it becomes obvious that in Germany – similar to the UK – the integration of EMAS on the national as well as the sub-national level is lacking. Thus, the effects of the policy instrument are primarily concentrated on direct environmental effects and remain internal while the work with indirect environmental issues would require larger policy integration. Consequently, the effects of EMAS are reduced to rather small-scale improvements while larger issues remain untouched by an instrument which could have much larger impacts on policy design especially in public authorities if it were integrated into a holistic policy concept and if the actors were willing to concentrate more on indirect effects rather than on small scale direct outcome reductions.

7.3 The EMAS policy process and its outcomes

Generally, the analysis of the EMAS policy processes revealed that continuous political support for the programme is vital for its development and success. Where it declined due to changes of political majority, the change of political priorities for policy programmes or as a result of financial problems of the authority, direct effects on the outcomes of the EMAS programmes are visible. Equally important is the motivational support provided by the leading actors of the policy process; the senior management team of an organisation as well as the EMAS coordinator are the leading figures. If these actors change their motivation, position or function, the policy programme is directly affected. Furthermore, they have a direct impact on the motivation of other staff to take part in the policy programme. Due to the primarily internal effects of EMAS on the operation of the organisation (in regard to its environmental outcomes), the senior management team in its executive function seems more important than the legislative structures of the organisation. Although the latter are important in the initial decision process as well as in regard to the policy continuation or termination, the ongoing operational work with the management system is done by the administration itself. This is due to the voluntary idea of the instrument and also reflects that the primary effects of EMAS are of internal nature to the organisation.

The analysis *of the agenda setting phase* has proved that there was a discrepancy between the mode of operation of the organisations and the political tendencies of the late 1990ies, when environmental issues were an important topic in both countries. Nevertheless, the use of management systems for public authorities was considered differently in the UK and Germany. In Britain, first the combination of NPM instruments and the introduction of environmental management systems, which would have been a useful combination, was not taken up by central government due to their political priorities. Nevertheless, some authorities joined EMAS with their other management instruments to achieve general and overall outcome improvements on environmental issues. In Germany, the organisations using EMAS had an incremental interest in changing their environmental outcomes; public authority reform processes or the reference to NPM-instruments was not the within the focus when introducing the European environmental standard. Further, in contrast to the UK authorities, the momentum of integration into the overall management system of the organisation has been achieved only by organisations that have gained a detailed understanding of the system because it required more changes to the operational structures of the organisations due to the lack of experience with NPM-instruments.

The *policy formulation* for this instrument is structured according to the EMAS regulation which provides a framework for the whole process to follow. In some cases, as with the authority UK 3, the existing environmental policy was well developed and required only minor adaptation to meet the EMAS standards. At other authorities like GE 3, a whole new policy field had to be developed; it was continued over the years with questionable results in the end. Thus, the policy formulation is dependent on the capabilities of the organisation. If it already has a good environmental policy record, an environmental management system is the frame for all activities. In this case, EMAS has the role of a top-up policy instrument which is in line with the premium environmental management system approach compared to other EMS. If, on the contrary, an organisation has a relatively low environmental profile and wants to use EMAS to improve on its environmental performance outcomes or to work on specific problems that it otherwise would not consider working on, the institutional framework is vital for a successful operation. As a consequence of both findings, an evaluation of the organisations' capabilities prior to the use of EMAS is absolutely necessary. Furthermore, reflections on the strategic goals that are to be achieved with the environmental management system are of great importance for the successful policy process.

The *decision making phase* for the primary introduction of the policy instrument was dominated by the senior management team within all organisations. While the reasons for introduction were connected to the political priorities of that time, the decision to continue or to terminate EMAS depended on the political and financial situation of the authority.

In the authorities of this study, the *implementation phase* of the environmental management system took about two years – from the first idea of the system to the first external validation and the registration with the relevant body. These processes were very dependent on the authority's staff and capabilities at the beginning of the introduction. Over the years, the protagonists of EMAS have gained considerable knowledge about the management system and its effects on the daily routines of officers, the general public and contractors. The organisations as a whole have learned how to interpret and use the EMAS regulation to the extent that, in some cases, it has even become a part of the management culture. Some authorities went further and integrated it into the overall management structures and systems of the organisation. Here, UK authorities have an advantage over the German ones because they already have management instruments in place that resemble the approach of EMAS. The implementation phase revealed that in all cases of this study, environmental management was connected to other projects, either before, during or after the introduction of EMAS. Often, these projects were a reason for using an EMS or they lead to changed views and activities concerning environmental issues after the authority had begun using an environmental management system.

The introduction of EMAS was not only done in order to improve the internal operation of the organisation. A major driver for introducing it was the suggested or implied changes of the image of the authority in the eyes of higher government authorities, the industry of the region and the general public. Nevertheless, all organisations claimed a minimal response to their EMS activities by stakeholders outside the authority (with the exception of a very small community of other authorities or interested researchers). Consequently, EMAS has to be seen in a larger context – it can stimulate an organisation to work more intensively on environmental issues or to cooperate in national or international projects. However, if there is no positive response (or even no response at all) from relevant actors to the fact that an organisation uses such a voluntary environmental instrument, a major driver is lacking, thus leading to a relatively low number of participants or even to a decline of the system. This is especially the case for local public authorities which are under the observation of the general public. Despite the differences comparing local public administration in the UK and Germany, the lack of incentives and positive feedback is a central reason for the declining interest for this policy instrument. In times where the financial situation of public administration is tight and a debate of the tasks of state is ongoing, the lack of support of a voluntary instrument makes it difficult for the organisation's executive to promote and continue with it.

A complex management system needs well trained staff to work with it. The main implementation work has to be done by the EMAS coordinator, who is permanent member of the authority. Where this is not the case, the organisations tend to buy capacity and knowledge from consultancies. This is in parts problematic because there is no constant work with the management system. What is more, the consultant usually does not have the close internal contacts and information from the other members of staff; consultants often only know the espoused theories but they lack knowledge of the theories in use of a special situation within the organisation. This way of working with the system can be seen as partly unsatisfactory because EMAS is not developed from within the organisation and not adapted to the needs of the users. Nevertheless, the temporally limited use of consultants is symptomatic for NPM-instruments. The results of the analysis have shown that usually, the EMAS coordinator works in close cooperation with a standing EMAS working group. This form of teamwork is essential for the implementation process because the information exchange from different departments and the knowledge base of the team members is strongly needed. Where there is no such working group, the cooperation of the different departments decreases and becomes difficult. Other than the traditional work within departments, the EMAS working group usually represents a horizontal working approach. In some organisations, this has led to follow-up projects, showing that this horizontal work on environmentally-related issues has a large potential when the relevant structures are given. In view of organisational change processes, these horizontal experiences with a focus on the subject rather than a straight-forward line structure can be a model for a changed organisational working structure.

Regarding the *continuation of the policy process*, all authorities expressed their commitment to continue with EMAS at the time of the interviews in the first half of 2005. Five have done so; at least one has definitely terminated the process. The reason for continuing the work is that an environmental management system is structurally designed as a long-term project; once authorities have decided to begin with it, they tend to stay with it for a longer period of time, especially if they are successful with it. Therefore, as the data has already indicated, the organisation GE 3, where EMAS depended to a large extent on outside input both financially as well as regarding the knowledge needed to continue the work, has terminated the policy programme. Although this was not only done due to a lack of resources, it is clear that the new organisational structure was a welcomed reason to terminate an already slowly-dying policy programme. Regarding the other authorities, there have been changes in the work with the management system over the years. One change is the integration of the system into the general management of the authority. If this is done in a cooperative and useful way, as at the authorities of UK 1, UK 2 and UK 3, this is a benefit for the whole organisation, although this can lead also to a reduction in the EMAS and/or the environmental activities in general because these issues lose importance through the integration. Another possible development is the reduction of EMAS activities because the easy (mostly direct environmental) targets have been achieved within a certain period of time. Then, the environmental management system is more or less a static system that is kept but not developed if it is not extended to other buildings or parts of the authority or if there are no new projects that are begun. This has been the case in the authorities of GE 1 and GE 2. Here, the general question of how long they will keep the system alive has to be asked. This element of a changed approach towards the policy instrument by its users over time has not been taken into account during the further development of the EMAS regulation by the European Commission because the Commission always assumes that the use of EMAS is ongoing at the same high level of activity and interest the organisations had when they started using it. Thus, one idea to keep the participants working with the regulation would be to allow a downsizing of the EMAS activities towards a set of minimum requirements. Furthermore, the idea of a useful policy termination should be a possible option which would, in turn, lead to new questions of the policy instrument's design and the relevant framework, especially if one considers that it is more useful to have authorities using the scheme for several years and then discontinuing the validation, for example, while maintaining a higher standard of environmental performance compared to the time when they had no EMS at all rather than having authorities with no structured environmental activities at all.

In view of the results of the analysis of the policy processes of the six organisations analysed within this study, one can summarise the outcomes by stressing the important role of the central actors in the policy process and their will to begin and continue with a long-term policy instrument. Where EMAS is seen as a method or tool to improve the organisation's own environmental policy processes, it can be successful. However, where it is seen as being a short-term instrument to improve the public image of an organisation or as a tool to organise additional funds for organisational reforms by using policy fields that are seen as highly important during a period of time, it will fail.

Regarding the environmental outcomes of the use of EMAS (which was, in quantitative terms, not within the focus of this study), it becomes clear that there are large differences between the organisations. The organisations were most active in the field of direct environmental effects, but the indirect effects were tackled only at the minimum. The outcomes of the direct effects also vary considerably. Therefore, regarding the future of EMAS, a discussion of minimum outcome standards is necessary, ass a larger focus on the indirect environmental effects that have to be taken into account. The direct environmental outcome could take the Japanese Top Runner programme, which defines efficiency goals for classes of products within a number of years (Jänicke 2008), as an example. This idea could be used within EMAS to define minimum goals for the environmental outcome for products, services and the operation of an organisation. While the flexibility of the use of EMAS would be reduced and the requirements for its use would be increased, the definition of some standard key figures in relation to the organisation would secure a minimum environmental outcome improvement and would add to the premium demand of the European regulation for environmental management.

7.4 Outcomes of organisational learning

The organisational learning is an integral part of processes of introducing, implementing and developing the environmental management system. The analysis verified the assumption that there are learning processes in every authority when using EMAS, although these differ in quantity, quality and regarding their long-term effects. On the topic of the espoused theories versus the theories-in-use, it becomes clear that all organisations see themselves as very environmentally aware and claim to have an improved image regarding environmental impacts and the responsibility for future generations. Nevertheless, the theories in use revealed that some organisations keep the system just to improve their environmental image (UK 1, GE 3) after some years of working with it while, at the same time, these authorities are internally questioning the activities carried out with and for the system. Other authorities are more proactive with EMAS and tend to extend and improve it rather than just keeping it (UK 2, UK 3), although they all have no clear indicators for the public awareness of the system. A third group of authorities within this study (GE 1 and GE 2) are active but tend to stay at their relatively high achieved level compared to the other authorities.

Further, all organisations see very low response rates to their environmental management activities from the general public as well as from the industry within the region. What is more, the regional or national policy authorities do not recognise the achievements with EMAS within their policy development, and even at the European level, there are no major active and positive recognitions for EMAS participants within EU legislation. This lack of policy integration on all three levels of politics is also reflected in the relevant legal norms. There are hardly any connections between EMAS and other environmental legislation, nor are there important benefits for authorities using EMAS, be it financially or legally. Additionally, although some organisations (UK 2 and GE 3) have actively participated or were asked to take part in European projects, there are no overall and clearly-visible legal and financial incentives on the European level for local authorities to take part in this European voluntary scheme and to present themselves with it. Without doubt, it is difficult to have the same standard of incentives within all EU member states regarding the EMAS regulation. Knowing this, the European Commission should lead the way and develop its own programmes to raise the attractiveness of participation in the system.

The different forms of learning – or loops – had an effect throughout the introduction period and while the work with EMAS was going on. The *single loop activities* are corrections of errors in order to maintain the central routines of the authority. In the case of using EMAS, these are primarily issues in the field of small technical improvements and development of procedures to reduce the water and energy use and creation of waste. All authorities have been active conducting such improvements. The success of this kind of activities depends on the targets that have been set as well as on the financial resources that have been used for these projects. In some cases, the outcomes of these activities are rather small compared to the work required to introduce and maintain a management system and sometimes financial restraints have reduced the achievable outcomes. Thus, the learning capabilities of organisations are able to manage single-loop changes because, primarily, they do not question the daily routines and methods of work as such. If, however, the EMAS processes remain at this level, the effectiveness of the system is questioned shortly after its introduction. This is one reason for a large number of organisations having terminated their work with the management system within a short time after introduction, not only in Germany but also in the UK. Therefore, in the revision of the EMAS regulation, a minimal target setting could improve the effectiveness of the system because it would give a more realistic view of what is expected by an organisation that uses EMAS and prevent authorities from terminating their work with the system shortly after its introduction. Additionally, the idea of working on easy targets first and then increasing the level of difficulty for the tasks to come during the work with EMAS could be an aspect to be considered in more detail in the external validation process.

The *double loop activities*, lead to larger changes within the organisation and show a much wider range of change processes and activities. First and foremost, all authorities have had to build up data for their environmental management system and for the documentation of it. Some authorities introduced a data management system, sometimes connected to the overall reporting system that is necessary according to the national rules. Here, the most active authorities (UK 2 and UK 3) indicated that a good data management system is a valuable advantage of EMAS in order to make changes and improvements visible. In contrast, authorities like the German GE 3 which had problems with data management seem to have problems generally with some parts of EMAS. This is especially true for organisations that lack the personnel and financial resources or for those which have no experience at all with management instruments. Comparing the data management capabilities of the authorities within this study, the British ones are performing better than their German counterparts due to the overall reporting requirements for Central government. As a consequence, one can conclude that the NPM-orientation of authorities supports their data management capabilities because these are already part of the modernisation concept. In this way, an NPM-orientated authority is predestined for EMAS compared to an organisation that is still run using the classical Weberian ideas of operation.

The majority of authorities reported an improvement of internal communication as a direct result of the work with EMAS. It is evident that the cross-departmental approach, which is a key element of the policy programme, is valuable especially for issues that are cooperatively worked on by several units of the authority. This is especially true for indirect environmental aspects. Both the communicative as well as the cross-departmental factor have to be strengthened in the future EMAS III regulation because it is fruitful for the whole management process. Compared to the German authorities, issue-working groups for the British authorities have long since been part of the management orientation. Especially for the German authorities, which tend to be more hierarchically structured than the British ones, this vertical form of cooperation is a productive requirement to tackle horizontal environmental issues. Furthermore, with the initiation of a permanent working group for

EMAS, the tasks of this policy area can be better organised and have a continuous working base. It becomes clear that without such a group, working with the system on a long-term basis is not possible.

The work with the regulation generally revealed a large amount of documentation that was sometimes seen as too detailed and demanding. Therefore, the organisations themselves began to reduce this documentation over the years, in line with the integration of EMAS into the general management system. This is parallel to the increase in flexibility regarding the work with the regulation in order to develop a more adaptable management system for the needs of staff. Over time, most of the organisations learned how to adapt the management system to their needs. This de-regulative as well as the adaptive nature of the management system are part of its strength. The trend towards an easier documentation has been going on for several years now and was one reason for the development of less-demanding environmental management systems (Bundesministerium für Umwelt, Naturschutz und Reaktorsicherheit 2005). Comparing the situation at the British and German public authorities, the latter ones had an advantage due financial aid, state initiatives and the development of the EMAS introduction guidelines given by federal and state government organisations. In contrast, the UK authorities had already gained expertise with documentation needs due to reform processes within the public sector.

Other effects of the double loop learning phase were an increase in the enthusiasm and support of the staff, which developed a culture of improvement, awareness and change in most authorities. It is not surprising that these effects were especially visible in the authorities that were very committed to EMAS. These authorities, therefore, were able to work on larger activities and changes than others. In line with this is the higher qualification of some of the coordinators at the authorities UK 3, GE 1 and GE 2. Their work with EMAS developed from other environmental policies and became a good basis for the broader view on environmental activities with the EMS. Consequently, the result is that incentive systems for policy programmes for the individual actor as well as for an organisation can be successful and should be considered whenever new programmes are designed and implemented, regardless of the organisation's mode of operation (Weberian or NPM-orientated).

Within the *deuteron learning phase*, a reflection of previous contexts is followed by a complete re-structuring of patterns of behaviour, strategies and targets of the organisation. These activities vary considerably in each authority. Overall learning activities can be seen at the majority of organisations that achieved a high level of integration and activity with the system. These organisations have learned to integrate the management system into their procedures and installed a constant process of improvement. They have also experienced the pressure that the regular (internal and external) validation puts on them. Although most organisations claimed that they could have done the activities under EMAS without the use of the policy instrument, they admitted at the same time that they would not have done so due to a lack of pressure. Also, structural deficits would have prevented them from working on environmental issues on a horizontal perspective. Only at one organisation, GE 3, were overall learning activities not found within this study.

All authorities reported follow-up projects of EMAS, covering a wide range of activities. The authorities developed numerous initiatives ranging from a large survey asking about the people's needs (not exclusively environmental issues) to the participation in European projects related to the environment. Major projects were the reconsideration of the energy strategy leading to investments in renewable energy production, projects to transfer the EMAS knowledge onto other organisations and the introduction of a facility management as well as a central procurement office. Although some of these projects were temporary, most of them were designed to have a long-term effect. It was the use of the management system that made things clear and visible and at the same time put pressure on the authorities and its leading figures to act. For those working with EMAS for several years now, it has become clear that a management framework can make a difference for public authorities that are often somewhat passive regarding changes of internal structures and routines. Consequently, the integration of EMAS as a regulated self-regulation (Jänicke 2008) into the overall management system of authorities as a way of promoting and implementing long-term environmental process innovation is an essential step in the development of the EMAS regulation. While maintaining its prime standard character, it is necessary to increase the policy instrument's integration in order to raise the number of organisations using the scheme in the future. To conclude, one can say that EMAS is designed to promote change and learning within the organisation using it. Where these processes are promoted and made possible, an environmental management system can have a positive effect. Where authorities are not able or willing to consider their modes of operation and the outcomes of them, such an instrument as EMAS is a mere tokenism to increase the organisation's external image.

7.5 Final Remarks

Regarding the idea of examining the introduction and implementation of EMAS within public administration, it first becomes clear that with all authorities using the scheme, EMAS has provided a structure around the political will of both the executive as well as the legislative. The work with the environmental management system has changed the culture of the organisations, although there are remarkable differences regarding the individual effects on the organisations. Where EMAS was implemented successfully, it has strengthened the organisation and led to the development of a long-term perspective on environmental issues, which is usually the result of an already environmentally-active authority. In contrast, at organisations where EMAS was not successfully implemented one can state that these authorities had not been active in environmental issues before introducing this horizontal instrument. The conclusion is that organisations that want to be successful with an environmental management system like EMAS already need a high environmental profile and level of activity. Organisations that want to introduce EMAS and do not have a high environmental profile need to develop many more organisational capabilities compared to the above-mentioned authorities. Further, with these organisations, it is more likely that they will fail with EMAS.

Whilst primarily the direct environmental aspects have been tackled by the organisations in detail, some authorities have shown capabilities and activities to work on indirect aspects. These activities have taken a longer time and have not been done to a great extent. With local public authorities especially, the work with indirect environmental aspects would have a large impact on the environmental policy. With planning, housing, traffic issues and local transport, much more could be achieved for an environmentally-friendly operation of the organisation than through the direct aspects, which are a good way of learning how to work on environmental effects but do not have such large effects compared to indirect aspects.

The authorities of the study have demonstrated considerable achievements in their environmental policies. Most of all, the success of EMAS is the result of long-term political support by central individuals of each organisation who integrated the system into the policy plan and programme already in place with the authority. In addition, the policy process of this instrument is important. If there are capable and willing actors, a clear defined political will, a focus on procedures and outcomes as well as a long-term commitment, there is a good chance that EMAS will be a success. The study has revealed that most authorities already active in environmental issues were actively looking for an overall system that would give a general structure to their different activities. In these cases, EMAS is the right instrument, framing the organisation's environmental efforts and supporting its future development within this policy field.

Regarding the differences of introducing and implementing EMAS in local public authorities in Germany and the UK in view of the developments described in chapter three, one can see that the British authorities have achieved a better integration of the system into their general management structure due to their NPM-orientation. The focus on performance indicators, a management-driven operation of the council as a whole as well as the outcome comparison of different councils under national standards have provided a good starting point for the introduction of an environmental management system. Despite this, the number of participants in EMAS throughout the UK indicates that the NPM-orientation is not the key reason for local authorities to start with environmental management. It is primarily the environmental awareness due to political, social and/or geographical reasons that provides a stable ground for local public organisations to question their own environmental outcome. Or to be more precise: the local policy arena is the key to a successful implementation of EMAS. For the European regulation on environmental management systems, the NPM structures in the UK are a positive knowledge base for the implementation and also for organisational learning. The fact that the outcomes of EMAS are not recognised within the national performance outcomes combined with a lower environmental awareness led to fewer local public authorities using the scheme.

As at the British authorities, the introduction of EMAS in German local public organisations depends on individuals that come up with the idea of using an environmental management system as a result of a policy discrepancy. However, in contrast to its UK counterparts, the EMAS activities within German authorities depend much more on the governing structure of an authority. Where it is strongly Weberian orientated, the horizontal approach of EMAS tends to work less successfully compared to authorities that have achieved a level of organisational modernity. In organisations which are still organised according to the classical principles of administration but have started the process of modernisation, the senior management team supports the long-term process of EMAS that is organised according to issues, topics and targets to be achieved rather than according to the hierarchical structure of the administration itself. Although EMAS is not understood as an NPM-instrument as such by the German authorities working with it, the system provides structures that introduce principles of NPM to the organisation. The numbers of participants in Germany indicate that over the years, EMAS has been more successful among Germany's public authorities on the regional and local level, primarily due to the support of the policy instrument by some states which took up the regulation in their environmental policy portfolio. Again, EMAS is successful where it connects to the sub-national environmental policy arena, but compared to the UK, it is a top-down support structure while the interest in the system is bottom-up. The German advantage in using EMAS is the higher environmental awareness compared to the UK, together with a larger flexibility regarding the operational principles of the organisation due to the federal structures, whereas the classical Weberian orientation of public administration tends to be more an impediment.

Generally, the NEPI idea of flexible and adaptable instruments is successful when there is a combined strategy of environmental policy that integrates EMAS. Yet, where EMAS has been introduced as a single instrument on its own, the system mainly has internal effects and is neglected by an overwhelming number of potential users. Therefore, EMAS is a classical example of a well-intentioned and well-designed policy instrument of active policy self-regulation that by far does not reach its full potential because of a lack of policy integration.

For the future of the EMAS regulation, a stronger recognition within EU legislation as well as within national and sub-national law is necessary to increase the attractiveness of the management system. Together with carefully-considered incentives, a reduced amount of necessary documentation and a large information campaign, the NEPI EMAS can still become a success. If the special needs of public administrations are recognised within the revised regulation, the interest and activities of this group of participants will increase again. This study has proved that the structural design of EMAS is working – it is the surrounding structure, i.e. the lack of policy integration as well as missing external incentives which are both critical for the success of the system. Thus, for the public authorities, EMAS is merely an internal management system. In this way, this type of organisation behaves in a similar way to industrial organisations (Clausen/Keil/Jungwirth 2002, IEFE 2005). The policy programme EMAS is not seen or used as a tool to improve the organisation's influence in the policy field of environment as a whole regarding its own operation or the activity of other actors. Public authorities are not using the knowledge and capabilities of the system to change their external policy behaviour. Consequently, EMAS loses much of its potential. If public authorities saw EMAS as a policy tool for the horizontal improvement of all their environmental activities, this would imply intensifying the efforts in the field of indirect environmental aspects as well as using the knowledge gained through working with the management system for strategic policy decisions outside the policy field of environment. This, in turn, would coincide with the ambitious ideas of authorities to be (environmentally) leading by example and to take responsibility for the future policy development.

7.6 Need for further research

The findings of this study demand further research which can be divided into two main areas. One is directly linked to the policy instrument EMAS and its use in different forms of organisations. While one large field of criticism about the EMAS regulation is that it is not specific to special needs of certain types of organisations, this analysis has shown that there are characteristics of public administrations that need to be taken into account in order to extend the potential of the policy programme EMAS. Therefore, a detailed analysis and comparison of the implementation processes of the instrument in different types of organisations like industry, service organisations, public authorities etc. would lead to organisation-type specifics of the EMAS process. These findings could be used to tune the regulation and thus improve its usability for different types of organisations. Regarding public authorities as a special type of organisation, a large, Europe-wide study that would examine the use of the regulation within public authorities could verify the result of this study that the national and regional characteristics of administrative governance strongly determine the use of EMAS within European countries and that the prime driver for the use of a voluntary instrument is the intrinsic motivation of the leading actors. The results of such a study could then lead to a more national or sub-national setting-specific adaptation of EU regulations on the one hand, and to a more detailed knowledge of the outcomes of policy instrument design on the EU level on the other.

A second area of further research could be the detailed development and testing of NEPIs. In view of the implementation deficits of numerous member states that are regularly documented by the European Commission, the basic idea of these instruments as being more flexible and adaptive to the specific needs of its users rather than continuing with the old command-and-control structure that does not seem appropriate as the EU expands, new instruments should be developed and evaluated further in order to see which influences NEPIs will have on the environmental policy mix and its outcome.

References

Altmann-Schevitz, J. et. al. 2002: European Conference: The EU Eco-Management and Audit Scheme. Benefits and Challenges of EMAS II. Final Report, Berlin 2002.

Andersen, B., Metcalfe, S. and Tether, B. 2000: Distributed Innovation Systems and Instituted Economic Process. In: Metcalfe, S. and Miles, I. (eds.): Innovation Systems in the Service Economy. Boston, Dordrecht: Kluwer, 15-42.

Anderson, J. E. 1975: Public Policymaking. 4th ed. New York: Houghton Mifflin.

ANEC et. al. (eds.) 2006: Joint ANEC/BEUC/ECOS/EEB position on Making EMAS a system of excellence – Going beyond EMS. Brussels (http://www.anec.org/attachments /ANEC-ENV-2006-G-047.pdf).

Argyris, C. and Schön, D. A. 1978: Organizational Learning: A theory of Action Perspective. Reading: Addison-Welsey.

Bacharach, S. B. and Lawler, E. J. 1981: Power and politics in organizations. San Francisco: Jossey-Bass.

Bandelow, N. 2003: Policy Lernen und politische Veränderungen. In: Schubert, K. and Bandelow, N. 2003: Lehrbuch der Politikfeldanalyse. München/Wien: Oldenbourg, 289-331.

Bandura, A. 1979: Sozial-kognitive Lerntheorie. Stuttgart: Klett-Cotta.

Banner, G. 1991: Von der Behörde zum Dienstleistungsunternehmen. In: Verwaltung, Organisation, Personal, 1, 5-12.

Barnes, P. 1994: The Environmental Audit Scheme. In: EIU European Trends, 3/1994, 84.

Barras, C. et. al. 1998: Transcriber: A Free Tool for Segmenting, Labeling and Transcribing Speech. First International Conference on Language Resources and Evaluation (http://trans.sourcefourge.net/articles/ transcriber-LREC11998.pdf, 23 April 2005).

Barras, C. et. al. 2000: Transcriber: development and use of a tool for assisting speech corpora production. Speech Communication special issue on Speech Annotation and Corpus Tools, Vol 33, No 1-2 (http://trans.sourcefourge.net/articles/Transcriber-SpeechComm 2000.ps, 23 April 2005).

Bartunek, J. M. 1984: Changing interpretative Schemes and organizational Restructuring: The Sample of a Religious Order. In: Administrative Science Quarterly, 29, 355-372.

Baumast, A. 1998: Die Entstehungsgeschichte des Umwelt-Audit. In: Doktoranden-Netzwerk Öko-Audit e. V. (eds.): Umweltmanagementsysteme zwischen Anspruch und Wirklichkeit. Berlin/Heidelberg: Springer, 33-56.

Baumast, A. 2003: Environmental Management Systems and Cultural Differences. An Explorative Study of Germany, Great Britain, and Sweden. Aachen: Shaker.

Bayerisches Landesamt für Umweltschutz (without year) (ed.): Umweltmanagement in Kommunen. Ein praktischer Wegweier für umweltbewusstes Handeln. Augsburg: Landesamt für Umweltschutz.

Bayerisches Staatsministerium für Landesentwicklung und Umweltfragen et. al. 2001 (eds.): EMAS – das neue EG-Öko-Audit in der Praxis. München.

Becker, B. 2002: Politik in Grossbritannien. Paderborn: Schöningh.

Benda, E. 1995: Rechtsstaat. In: Nohlen, D. (ed.): Lexikon der Politik. Band 1. München: Beck, 515-519.

Bentlage, J. 1999: Potentiale der Entwicklung und Implementierung eines Umweltmanagementsystems im öffentlichen Dienstleistungsbereich. Erlangen/Nürnberg.

Benz, A. et. al. 2003: Governance. Eine Einführung. Kurs 3202 der FernUniversität Hagen. 9. Auflage. Hagen: Fernuniversität. Bertelsmann Stiftung 1993 (ed.): Demokratie und Effizienz in der Kommunalverwaltung. Band 1. Gütersloh: Bertelsmann-Stiftung.

Benz, A. et. al. 2007: Handbuch Governance. Theoretische Grundlagen und empirische Handlungsfelder. Wiesbaden: VS, Verlag für Sozialwissenschaften.

Bertelsmann Stiftung (ed.) 1993: Carl Bertelsmann-Preis. Demokratie und Effizienz in der Kommunalverwaltung. Band 1. Dokumentationsband zur internationalen Recherche. Gütersloh: Verlag Bertelsmann Stiftung.

Blanke, B. et. al. (eds.) 2001: Handbuch Verwaltungsreform. 2nd ed. Opladen: Leske und Budrich.

Blanke, B. et. al. (eds.) 2003: Modernes Management für die Verwaltung - ein Handbuch. Hannover: Pinkvoss.

BMU/UBA 2001: Handbuch Umweltcontrolling für die öffentliche Hand. München: Vahlen.

Böhret, C. 2001: Verwaltungspolitik als Führungsauftrag. In: Blanke, B. et. al. 2001 (eds.): Handbuch Verwaltungsreform. 2nd ed. Opladen: Leske und Budrich, 43-49.

Bogumil J. 2002: Zum Verhältnis von Politik und Verwaltungswissenschaft in Deutschland. polis Nr. 54 / 2002 (http://www.fernuni-hagen.de/POLAD/download/polis54.pdf).

Bogumil, J. et. al. 2005: Zehn Jahre Neues Steuerungsmodell. Eine Bilanz kommunaler Verwaltungsmodernisierung. Berlin: edition sigma.

Bogumil, J. and Jann, W. 2005: Verwaltung und Verwaltungswissenschaft in Deutschland. Einführung in die Verwaltungswissenschaft. Wiesbaden: VS Verlag.

Bogumil, J. and Kuhlmann, S: 2004: Zehn Jahre kommunale Verwaltungsmodernisierung – Ansätze einer Wirkungsanalyse. In: Jann, W, et al.: Statusreport Verwaltungsreform – Eine Zwischenbilanz nach 10 Jahren. Berlin: Edition Sigma.

Borins, S. and Grüning, G. 1998: New Public Management Theoretische Grundlagen und problematische Aspekte der Kritik. In: Budäus, D., Conrad, P. and Schreyögg, G. (eds.): New Public Management. Berlin: De Gruyter, 11-54.

Bouma, J. J. 2000: Environmental management systems and audits as alternative environmental policy instruments? In: Knill, C. and Lenschow, A. (eds.): Implementing EU environmental policy. New directions and old problems. Manchester: Manchester University Press, 116-133.

Bower, G. H. and Hilgard, E. E.1983: Theorie des Lernens I. 5th ed. Stuttgart: Klett.

Boyne, G. A. 1999: Introduction: Processes, performance and best value in local government. In: Local Government Studies, 25: 2, 1-15.

Bracke, R. and Albrecht, J. 2007: Competing environmental management standards: how ISO 14001 outnumbered EMAS in Germany, the UK, France and Sweden: In: Environment and Planning C: Government and Policy 25(4), 611-627.

Brewer, G. 1974: The Policy Sciences Emerge. To Nurture and Structure a Discipline. In: Policy Sciences, 5, 239-244.

Buchanan, J. M. and Tullock, G. 1962: The calculus of consent. Logical foundations of constitutional democracy. Ann Arbor.

Buchanan, J. M. 2003: Public Choice: The origins and development of a research programme. Fairfax (http://www.gmu.edu/centers/publicchoice/pdf% 20links/Booklet.pdf, 01 April 2007).

Bültmann, A. and Wätzold, F. 2000 : The implementation of national and European environmental legislation in Germany. Three case studies. UFZ research report 20/2000. Leipzig : UFZ.

Bulletin EU, http://europa.eu/bulletin/en/200011/p104037.htm.

Bulletin EU 1/2-2001, http://europa.eu/bulletin/en/200101/p104048.htm.

Bundesministerium des Innern 2005: Fortschrittsbericht 2005 des Regierungsprogramms „Moderner Staat – Moderne Verwaltung“ im Bereich des

Verwaltungsmanagements, Berlin. (http://www.verwaltung-innovativ.de/cln_046/nn_684264/SharedDocs/Publikationen/Bestell service/fortschrittsbericht__2005,templateId=raw,property=publicatio nFile.pdf/fortschrittsbericht_2005.pdf, 08 February 2008).

Bundesministerium des Innern 2006: Zukunftsorientierte Verwaltung durch Innovation, Berlin. (http://www.verwaltung-innovativ.de/cln_046/nn_684264/SharedDocs/Publikationen/Bestellservice/programm__zukun ftsorientierte__verwaltung,templateId=raw,property=publicationFile.p df/programm_zukunftsorientierte_verwaltung.pdf, 08 February 2008).

Bundesministerium für Umwelt, Naturschutz und Reaktorsicherheit/ Umweltbundesamt (ed.) 2005: Umweltmanagementansätze in Deutschland. Berlin/Dessau: Umweltbundesamt (http://www.bmu.de /files/wirtschaft_und_umwelt/emas/application/pdf/broschuere_um weltmanagementansaetze.pdf, 22 March 2007).

Bundesregierung/Kabinettsbeschluss 01.Dezember 1999: Moderner Staat – Moderne Verwaltung. Leitbild und Programm der Bundesregierung. (http://www.verwaltung-innovativ.de/cln_046/nn_684264/SharedDocs/Publikationen/Bestells ervice/programm__der__bundesregierung,templateId=raw,property=p ublicationFile.pdf/programm_der_bundesregierung.pdf, 08 February 2008).

Bundesministerium der Justiz 1995: Gesetz zur Ausführung der Verordnung (EG) Nr. 761/2001 des Europäischen Parlaments und des Rates vom 19. März 2001 über die freiwillige Beteiligung von Organisationen an einem Gemeinschaftssystem für das Umweltmanagement und die Umweltbetriebsprüfung (EMAS) (Umweltauditgesetz - UAG) (http://bundesrecht.juris.de/uag/BJNR159100995.html, 23 March 2008).

Bund-/Länder-Arbeitskreis steuerliche und wirtschaftliche Fragen des Umweltzschutzes (ed.): Bericht: EMAS – Sachstand und Bewertung. (without year, place and author) (http://www.blag-ne.de/dateien/dat_nr288_1.pdf, 20 February 2008).

Bungarten, H. 1976: Die Umweltpolitik der Europäischen Gemeinschaft. Bonn: Europa-Union Verlag.

Caspari, S. 1995: Die Umweltpolitik der Europäischen Gemeinschaft. Eine Analyse am Beispiel der Luftreinhaltepolitik. Baden-Baden: Nomos.

Chandler, A. D. 1962: Strategy and Structure: chapters in the history of the environmental enterprise. Cambridge, Mass: MIT Press.

Clausen, J. and Jungwirth, M. 2002: EMAS in öffentlichen Verwaltungen. In: Umwelt kommunale Ökologische Briefe, 25-26/02 26-27.

Clausen, J., Keil, M. and Jungwirth, M. 2002: The State of EMAS in the EU. Berlin.

Cohen, M., March, J. and Olsen, J. 1972: A Garbage Can Model of Organizational Choice. In: Administrative Science Quarterly Vol. 17, March, 1-25.

Copus, C. 2001: Local Government. In: Jones, B., et. al. (eds.): Politics UK. 4th ed. Harlow: Longman, 478-498.

Cyert, R. M. and March, J. G. 1963: A Behavioral Theory of the Firm. Englewood Cliffs, N. J.: Prentice-Hall.

Czada, R. 1998: Neuere Entwicklungen der Politikfeldanalyse. In: Alemann, U. von and Czada, R. (eds.): Kongressbeiträge zur politischen Soziologie, politischen Ökonomie und Politikfeldanalyse. polis, 39, 47-65.

Damkowski, W. and Precht, C. 1995: Public Management. Neuere Steuerungskonzepte für den öffentlichen Sektor. Stuttgart: Kohlhammer.

Davidow, W. H. and Malone, M. S.1992: The Virtual Corporation: structuring and revitalising the corporation for the 21. century. New York: Harper Collins

Delgado, L. 2000: The impact of the EU Eco-Audit regulation on innovation in Europe, Seville: Institute for prospective technological studies Sevilla.

Delogu, B. 1992: The EC Proposal on Environmetnal Auditing. In: Tagungsband zum Seminar: Die Praxis des Umweltaudits, Düsseldorf 27.05.1992. ICC-Publikation 210/4, 29-49.

Delmas, M.A. 2002: The diffusion of environmental standards in Europe and the United States: An international perspective. In: Policy Sciences 35: 91-119.

Deming, W. E. 1986: Out of the Crisis. 5th ed. Cambridge, Mass.: Massachusetts Institute of Technology, Center for Advanced Engineering Study.

Deming, W. E. 1993: Out of the crisis. Cambridge, Mass: MIT Press.

Demmke, C. 1994: Die Implementation von EG-Umweltpolitik in den Mitgliedstaaten. Umsetzung und Vollzug der Trinkwasserrichtlinie. Baden-Baden: Nomos.

Department of the Environment 1997: This Common Inheritance. UK Annual Report 1997. CM 3556. London: The Stationery Office.

Department of the Environment, Transport and the Regions 1998: Improving Local Services Through Best Value. London: HMSO. (http://www.local.odpm.gov.uk/bv/improvbv.improvbv.pdf, 03 June 2008).

Department for Environment, Food and Rural Affairs 2006 (eds.): Environmental Key Performance Indicators. Reporting Guidelines for UK Business. London: Queen's Printer and Controller. (http://www.defra.gov.uk/environment/business/envrp/envkpi-guidelines.pdf, 12 March 2008).

Deutscher Bundesrat: Bundesratsdrucksache 222/92: Vorschlag einer Verordnung (EWG) des Rates, die die freiwillige Beteiligung gewerblicher Unternehmen an einem gemeinschaftlichen Öko-Audit-System ermöglicht.

Deutscher Naturschutzring 1999 (ed.): Stellungnahme zu EMAS II. Bonn.

DHV Environment and Infrastructure BV 2002 (eds.): The Guide to Implement EMAS in the CEE countries. Amersfoort (http://europa.eu.int/comm/environment/emas/pdf/activities/ emas_guide.pdf).

Dosi, G. 1997: Opportunities, Incentives and the Collective Pattern of Technological Change. In: The Economic Journal, 107, 1530-1547.

Downe, J. and Martin, S. 2006: Joined up policy in practice? The coherence and impacts of the local government modernisation agenda. In: Local Government Studies, 32:4, 465-488.

Duncan, R. B. and Weiss, A. 1979: Organizational Learning: Implications for Organizational Design. In: Staw, B. (ed.): Research in Organizational Behavior. Vol. 1, Greenwich: JAI Press, 75-123.

Dye, T. R. 1976: Policy Analysis. What Governments Do, Why They Do It, and What Difference it Makes. University of Alabama: University of Alabama Press 1976.

Eichhorn, P. et. al. (eds.) 2002: Verwaltungslexikon. 3rd ed. Baden-Baden: Nomos.

Elcock, H. 1991: Change and Decay? Public Administration in the 1990ies. London: Longman.

Ellwein, T. 1966: Einführung in die Regierungs- und Verwaltungslehre. Stuttgart: Kohlhammer.

Ellwein, T. and Hesse, J. J. 1985: Verwaltungsvereinfachung und Verwaltungspolitik. Baden-Baden: Nomos.

Ellwein, T. 1993: Der Staat als Zufall und als Notwendigkeit. Die Jüngere Verwaltungsentwicklung am Beispiel Ostwestfalens. Band 1. Opladen: Westdeutscher Verlag.

Ellwein, T. 1994: Das Dilemma der Verwaltungsstruktur und Verwaltungsreform in Deutschland. Mannheim: BI-Taschenbuchverlag.

EMAS Helpdesk 2007: Report on the Public Consultation Revision of the EU Eco-Management and Audit Scheme. (EMAS) Regulation (EC) No 761/2001 (http://ec.europa.eu/environment/emas/pdf/news/ stake holder_consultation_report_2007_final.pdf).

EMAS Helpdesk (eds.) 2008: EMAS statistics, as of 27/11/2008 (http:// www.ec.europa.eu/environment/emas/pdf/5_5articles_en.pdf), 20 December 2008).

Emilsson, S. and Hjelm, O. 2004: Different approaches to standardized environmental management systems in local authorities – two case studies in Gothenburg and Newcastle. In: Corporate Social Responsibility and Environmental Management 11, 48-60.

Endruweit, G. 2004: Organisationssoziologie. 2nd ed. Stuttgart: Lucius und Lucius.

Environment Agency 2006: Results of the Remas Project (http:// publicati ons.environment-agency.gov.uk/epages/eapublications.storefront/485 827ef00da575a273fc0a80296065e/Product/View/SCHO0408BNXL& 2DE&2DE, 23 January 2008).

EU Official Journal (OJ) 1992: OPINION OF THE ECONOMIC AND SOCIAL COMMITTEE on the proposal for a Council Regulation (EEC) allowing voluntary participation by companies in the industrial sector in a Community eco-audit scheme. OJ C 332, 16.12.1992, 44.

EU Official Journal (OJ) 1993: OJ LEGISLATIVE RESOLUTION embodying the opinion of the European Parliament on the Commission proposal for a Council regulation allowing voluntary participation by companies in the industrial sector in a Community Eco-audit scheme. OJ C 42, 15.2.1993, 60.).

European Commission 1993: Council Regulation (EEC) No 1836/93 of 29 June 1993 allowing voluntary participation by companies in the industrial sector in a Community eco-management and audit scheme. OJ L 168, 10.7.1993. Brussels (http://eur-lex.europa.eu/LexUriServ/ LexUriServ.do?uri=CELEX:31993R1836:EN:HTML).

European Commission 1996: Communication from the Commission on Environmental Agreements (COM(96)561 final). Brussels.

European Commission 1996a: Commission Decision of 2 February 1996 on the recognition of the British standard BS7750: 1994, establishing

specifications for environmental management systems, in accordance with Article 12 of Council Regulation (EEC) No 1836/93 (96/150/EC). Brussels (http://eur-lex.europa.eu/LexUriServ/LexUriServ.do?uri=CELEX:31996D0150:EN:HTML).
European Commission 1996: Thirteenth Annual Report on Monitoring the Application of Community Law. Luxembourg: Office for Official Publ. of the Europ. Communities.
European Commission (ed.) 1998: Proposal for a Council Regulation (EC) allowing voluntary participation by organisations in a Community eco management and audit scheme ((COM 1998/622) Brussels. http://eur-lex.europa.eu/LexUriServ/site/en/oj/1998/c_400/c_40019981222en00070025.pdf).
European Commission (ed.) 2000: Opinion of the Commission pursuant to Article 251(2) (c) of the EC Treaty, on the European Parliament's amendments to the Council's common position regarding the proposal for a Regulation of the European Parliament and of the Council allowing voluntary participation by organisations in a Community eco-management and audit scheme (EMAS). (COM/2000/0512 final, http://eur-lex.europa.eu/LexUriServ/LexUriServ.do?uri=COM:2000:0512:FIN:DE:PDF).
European Commission (ed.) 2001: Regulation (EC) No 761/2001 of the European Parliament and of the Council of 19 March 2001 allowing voluntary participation by organisations in a Community eco-management and audit scheme (EMAS). Brussels. (http://eurlex.europa.eu/JOHtml.do?uri=OJ:L:2001:114:SOM:EN:HTML).
European Commission (ed.) 2004: Annex to the Report from the Commission to the Council and the European Parliament on Incentives for EMAS registered organisations. (COM(2004)745 final). Brussels. (http://www.ec.europa.eu/environment/emas/pdf/news/incentives_en.pdf, 25 March 2007).
European Commission, DG Environment, Neil Hamon 2006: The European Eco Management and Audit Scheme. Powerpoint-Presentation at the EMAS revision workshop, Brussels 10.-11. December 2006. Brussels.
European Commission 2007: 24th Annual Report on Monitoring the Application of Community Law. (COM(2007) 398 final).
European Commission, DG Environment 2007 (ed.): EMAS member state activities. Brussels (http://ec.europa.eu/environment/emas/activities/index_en.htm, 8 April 2007).
European Commission, DG Environment 2008 (ed.): Proposal for a Regulation of the European Parliament and of the Council on the

voluntary participation by organisations in a Community eco-management and audit scheme (EMAS). Brussels.(http://ec.europa.eu/environment/emas/pdf/com_2008_402_draft.pdf, 18 July 2008).

Franke, J. and Wätzold, F. 1995: Political Evolution of EMAS: Perspectives from the EU, nationalGovernments and Industrial Groups. In: European Environment, 5, 155-159.

Freier, I. 2005: Regional policy for diffusing environmental management systems from an innovation system perspective. Vechta: Hochschule Vechta.

Friedlander, F. 1983: Patterns of individual and organisational Learning. In: Shrivastava, S. (ed.): The Executive Mind. New Insights on Managerial Thought and Action. San Francisco: Jossey-Bass, 192-220.

Fronek, A. 2003: Umweltmanagementsysteme auf dem Prüfstand. Die Bedeutung von EMAS und ISO 14001 im Konzert der umweltpolitischen Systeme. Hamburg: Kovac.

Funk, R. 2000: Umweltmanagement und Umweltaudit für Kommunen. Aachen: Shaker.

Gagné, R. M. 1970: Die Bedingungen des menschlichen Lernens. 2nd ed. Hannover: Schroedel.

Geibel, J. 2004: Das EG-Öko-Audit als Instrument für einen umweltgerechten Güterverkehr? Hamburg: Kovac.

Gerstlbergerger, W. and Kneisller, T: 2000 Reformeriung von Verwaltung im Kontext der Verwaltungsmodernisierung. In: Zielinski, H. (ed.): Die Moerninsierung der Städte: Verwaltung und Politik zwischen Ökonomisierung und Innovation. Wiesbaden: Deutscher Universitäts-Verlag.

Geus, A. de 1988: Planning as Learning. In: Harvard Business Review. Vol. 66, 2, 70-74.

Glaser, B.G. and Strauss, A. L. 1967: The Discovery of Grounded Theory: Strategies for Qualitative Research. Chicago: Aldine Publishing Company.

Glaser, B. G. and Strauss, A. L.: 1998: Grounded Theory. Strategien qualitativer Forschung. Bern: Huber.

Giddens, A. 1997: The Third Way. Cambridge: Polity Press.

Gläser, J. and Laudel, G. 1999: Theoriegeleitete Textanalyse. Das Potential einer variablenorientierten qualitativen Inhaltsanalyse. WZB Discussion Paper P99-401, Berlin 1999. (http://skylla.wzb.eu/pdf/1999/p99-401.pdf, 17 February 2006).

Gläser, J. and Laudel G. 2004: Experteninterviews und qualitative Inhaltsanalyse. Wiesbaden: VS Verlag für Sozialwissenschaften.

Goldsmith, B. E. 2003: Imitation in International Relations: Analogies, Vicarious Learning and Foreign Policy. International Interactions. 29: 273-267.

Grahl, B. and Falkenberg, D. (eds.) 2001: Nutzung der Systematik von Umweltmanagementsystemen in Kommunen. Rahmenkonzept zur Verknüpfung kommunaler Handlungsebenen. Lübeck: Fachhochschule Lübeck.

Grauhan, R. 1969: Modelle politischer Verwaltungsführung. In: Politische Vierteljahresschrift, Vol 2/3, 269-284.

Greenwood, J. and Eggins, H. 1995: Shifting Sands: Teaching Public administration in a Climate of change. In: Public Administration. Vol. 73, 143-164.

Greenwood, J., Pyper, R. and Wilson, D. 2002: New Public Administration in Britain. London/New York: Routledge.

Greven, M. 2008: „Politik" als Problemlösung und als vernachlässigte Problemursache. Anmerkungen zur Policy Forschung. In: Janning, F. and Toens, K. (eds.): Die Zukunft der Policy-Forschung: Theorien, Methoden, Anwendungen. Wiesbaden: VS Verlag für Sozialwissenschaften, 23-33.

Grochla, E. 1969 (ed.): Handwörterbuch der Organisation. Stuttgart: Poeschel.

Grömig, E. and Gruner, K. 1998: Reform in den Rathäusern. In: Deutscher Städtetag (ed.): der städtetag, Heft 8, 581-587.

Grössmann, U. 2006: Betriebsbeauftragte für Umweltschutz. Konstanz. (http://www.umwelt-consulting-groessmann.de/veroeffentlichungen/betriebsbeauftragter.pdf, 23 June 2008).

Große, H. 2003: Anforderung von Umweltmanagementsystemen nach der EMAS-VO und der ISO 14001. In: Kramer, M., Brauweiler, J. and Helling, K. (eds.): Internationales Umweltmanagement. Band II. Umweltmanagementinstrumente und –systeme. Wiesbaden: Gabler, 135-194.

Guckenbiehl, H. L. 2000: Institution und Organisation. In: Korte, H. and Schäfers, B. (eds.): Einführung in die Hauptbegriffe der Soziologie. 5th ed. Opladen: Leske und Budrich.

Handel, M. J. (ed.) 2003: The Sociology of Organisations. Classic, contemporary, and critical readings. Thousand Oaks: Sage. Hedberg, B. 1981: How Organizations Learn and Unlearn. In: Nyström, P. C. and

Starbuck, W. H. (eds.): Handbook of Organizational Design. Vol. 1: Adapting Organizations to their Environment. Oxford: Oxford University Press, 3-27.

Heide, H.-J. von der 1998: Stellung und Funktion der Kreise. In: Wollmann, H. and Roth, R. (eds.): Kommunalpolitik. Politisches Handeln in den Gemeinden. Bonn: Bundeszentrale für politische Bildung, 123-132.

Heinelt, H. et. al. 2000: Prozedurale Umweltpolitik der EU. Umweltverträglichkeitsprüfungen und Öko-Audits im Vergleich. Opladen: Leske und Budrich.

Hellenthal, F. 2001: Umweltmanagement nach der Öko-Audit-Verordnung. Kritische Betrachtung und Darlegung von Perspektiven durch das Konzept der ökologischen Unternehmensbewertung. Marburg: Tectum Verlag.

Hennessy, P. 1998: The Blair style of government: an historical perspective and an internal audit. In: Government and Opposition, 33, 2-20.

Heritier, A. et. al. 1994: Die Veränderung von Staatlichkeit in Europa. Ein regulativer Wettbewerb: Deutschland, Frankreich, Großbritannien. Opladen: Leske und Budrich.

Herrmann, S. and Boguslawski, A. von 1997: Das hessische Pilotprojekt zum kommunalen Öko-Audit. In: Umwelt kommunale ökologische Briefe, 22, 8-12.

Hesse, J. J. 1990: Local Government in a federal state: the case of West Germany. In: Hesse, J. J. (ed.): Local government and urban affairs in International Perspective. Baden-Baden: Nomos.

Hesse, J. J. and Sharpe, L. J. 1991: Local Government in International Perspective: Some Comparative Observations. In: Hesse, J. J.: Local Government and Urban Affairs in International Perspective. Analyses of Twenty Western Industrialised Countries. Baden-Baden: Nomos, 603 -621.

Hesse, K. 1962: Der unitarische Bundesstaat. Karlsruhe: C. F. Müller.

Hildebrandt, E. and Schmidt, E. 1994: Umweltschutz und Arbeitsbeziehungen in Europa. Berlin: edition sigma.

HMSO 1999: White paper "Modernising Government". (http://archive. cabinetoffice.gov.uk/moderngov/download/modgov.pdf, 20 Juli 2006).

Hogwood, B. 1992: Trends in British Public Policy. Buckingham: Open University Press.

Holzinger, K. 1994: Politik des kleinsten gemeinsamen Nenners? Umweltpolitische Entscheidungsprozesse in der EG am Beispiel der Einführung des Katalysatorautos. Berlin: edition sigma.

Holzinger, K. 1995: Ökonomische Theorien der Politik. In: Nohlen, D. (ed.): Lexikon der Politik. Band 1: Politische Theorien. München: Beck, 383-391.

Holzinger, K., Jörgens, H. and Knill, C. (eds.) 2007: Transfer, Diffusion und Konvergenz von Politiken: PVS Sondergheft 38/2007.

Howlett, M. and Ramesh, M. 2003: Studying Public Policy. Policy Cycles and Policy Subsystems. 2nd ed. Don Mills, Ontario: Oxford University Press.

IEFE 2005 (ed.): EVER study Report 1. Options and Recommendations for the Revision Process. Milano. (http://ec.europa.eu/environment/emas/pdf/ everfinalreport1_en.pdf, 11 December 2005).

Isaac-Henry, K., Painter, C. and Barnes, C. 1997: Management in the Public Sector. Challenge and change. London: Chapham and Hall 1996.

Jachtenfuchs, M 1996: International policy making as a learning process? The European Union and the greenhouse effect. Aldershot: Avebury. Jacobs, M. and Levett, R. 1993 (For the Department of the Environment) (ed.): A Guide to the Eco-Management and Audit Scheme for UK Local Government. London: HMSO.

Jänicke, M., Kunig, P. and Stitzel, M. 1999: Lern- und Arbeitsbuch Umweltpolitik. Politik, Recht und Management des Umweltschutzes in Staat und Unternehmen. Bonn: Dietz.

Jänicke, M. 2008: Megatrend Umweltinnovation. Zur ökologischen Modernisierung von Wirtschaft und Staat. München: ökom.

Jann, W. 1986: Politikwissenschaftliche Verwaltungsforschung. In: Beyme, K. von (ed.): Politikwissenschaft in der Bundesrepublik Deutschland. Entwicklungsprobleme einer Disziplin. Politische Vierteljahresschrift. Sonderheft 17. Wiesbaden 1986, 209-230.

Jann, W. 1998: Neues Steuerungsmodell. In: Bademer, S. von et. al: (eds.): Handbuch zur Verwaltungsreform. Opladen: Leske und Budrich, 70-80.

Jann, W. 1991: Politikfeldanalyse. In: Nohlen, D. (ed.): Wörterbuch Staat und Politik. Müchen: Piper, 499-503.

Jann, W. 2002: Der Wandel verwaltungspolitischer Leitbilder: Vom Management zu Governance? In: König, K. (ed.): Deutsche Verwaltung an der Wende zum 21. Jahrhundert. Baden-Baden: Nomos, 279-303.

Jann, W. and Wegrich, K. 2003: Phasenmodelle und Politikprozesse: Der Policy Cycle. In: Schubert, K. and Bandelow, N. C. (eds.): Lehrbuch Politikfeldanalyse. München/Wien: Oldenbourg, 71-106.

Jann, W. and Wegrich, K. 2007: Theories oft he Policy Cycle. In: Fischer, F, Miller, G. J. and Sidnely M. S. (eds.) 2007: Handbook of Public Policy Analysis: Theory Politics and Methods. Boca Raton: CRC Press, 43-63.

Jann, W. and Wewer, G. 1998: Helmut Kohl und der „schlanke Staat". Eine verwaltungspolitische Bilanz. In: Wewer, G. (ed.): Bilanz der Ära Kohl. Opladen: Leske und Budrich, 229-266.

Jann W. et. al. (eds.) 2004: Status-Report Verwaltungsreform – Eine Zwischenbilanz nach 10 Jahren. Berlin: edition sigma.

Janning, F. and Toens, K. (eds.) 2008: Die Zukunft der Policy-Forschung. Wiesbaden: VS Verlag für Sozialwissenschaften.

Jelinek, M. 1979: Institutionalizing Innovation. A Study of Organizational Learning. New York: Praeger.

Jenkins, W. I. 1978: Policy Analysis. A Political and Organisational Perspective. London: Robertson.

Jessop, B. 1994: The transition to post-Fordism and the Schumpeterian workfare state. In: Burrows, R. and Loader, B.: Towards a Post-Fordist Welfare State. London: Routledge, p. 13-37.

Jones, B. et. al. 2001 (eds.): Politics UK. 4th ed. Harlow: Longman.

Jordan, A. 1999: Subsidiarity and environmental policy: which level of government should do what in the European Union? CSERGE Working Paper 99-13. Norwich: GEC.

Jordan, A. 2002: The Europeanisation of Pritish Environmental Policy.: A Departmental Perspektive. Basingstoke: Earthscan.

Jordan, A. Wurzel, R. and Zito R. (eds.) 2003: New Instruments of Environmental Governance. London: Frank Cass.

Kähler, M. and Rotheroe N. C. 1999: Comparison of the British and German approach towards the European Eco-Management and Audit Scheme (EMAS). In: Eco Management and Auditing, 6, 115-127.

Karrdorf, E. von 2003: Qualitative Evaluationsforschung, in: Flick, U., Kardorff, E. von and Steinke, I. (eds.): Qualtitative Forschung. Ein Handbuch. Reinbek: Rowohlt, 238- 250.

Kelle, U. 1995: Computer-aided qualitative data analysis. Theory, Methods and Practice. London: Sage.

Kelle, U. 2003: Computergestützte Analyse qualitativer Daten. In: Flick, U. et. al. (eds.): Qualitative Forschung. Ein Handbuch. Zweite Auflage. Reinbek bei Hamburg: Rowohlt Taschenbuch-Verlag, 485-502.

Kelley, M. J. 2004: Qualitative evaluation research. In: Seale et.al. (eds.): Qualitative Research Practice. London: Sage, 521-535.

Kern, K. et. al. 2001: Policy Convergence and Policy Diffusion by Governmental and Non-Governmental Institutions – An internatnonal Comparison of Eco-labeling systems. Discussion Paper FS II 01 - 305, Wissenschaftszentrum Berlin für Sozialforschung, http://skylla.wz-berlin.de/pdf/2001/ii01-305.pdf (February 22, 2008).

KGSt 1993 (ed.): Das Neue Steuerungsmodell. Bericht Nr. 5. Köln: KGSt.

KGSt 1994 (ed.): Das neue Steuerungsmodell. Köln: KGSt.

Kickert, W. and Jorgensen, T. B. 1995: Introduction: Managerial Reform Trends in Western Europe. In: International Review of Administrative Sciences. Vol. 61, 499-510.

Kingdon, J. W. 1995: Agendas, Alternatives and Public Policies. 2nd ed. New York: Harper Collins College Publishers.

Kirsch, W. 1990: Unternehmenspolitik und strategische Unternehmensführung. München: Kirsch.

Kißler, L. et. al. 1997: Moderne Zeiten im Rathaus? Reform der Kommunalverwaltungen auf dem Prüfstand der Praxis. Berlin: edition sigma.

Kleinfeld, R. 1996: Kommunalpolitik. Eine problemorientierte Einführung. Opladen: Leske und Budrich.

Knill, C. and Lenschow, A. 2000: Do new brooms really sweep cleaner? Implementation of new instruments in EU environmental policy. In: Knill, C. and Lenschow, A. (eds.): Implementing EU environmental policy. New directions and old problems. Manchester: Manchester University Press, 251-286.

Knill, C. and Lenschow, A. 2000a: Introduction: New approaches to reach effective implementation – political rhetoric or sound concepts? In: Knill, C. and Lenschow, A. (eds.): Implementing EU environmental policy. New directions and old problems. Manchester: Manchester University Press, 3-8.

Knill, C. 2003: Europäische Umweltpolitik. Steuerungsprobleme und Regulierungsmuster im Mehrebenensystem. Opladen: Leske und Budrich.

König, K. and Füchtner, N. 1998: Von der Verwaltungsmodernisierung zur Verwaltungsreform. In: König, K. and Füchtner, N. (eds.): Schlanker Staat. Verwaltungsmodernisierung im Bund. Speyer: Deutsche Hochschule für Verwaltungswissenschaften, 5-49.

König, K. and Miller, M. 1995: Vorstudien zur Organisation und Reform von Landesverwaltungen. Band 1 und 2. Speyerer Forschungsberichte. Speyer: Forschungsinstitut für öffentliche Verwaltung.

Kolb, D. 1984: Experimental learning experience as the source of learning and development. Englewood Cliffs, NJ: Prentice-Hall.

Kroeber-Riel, W. 1990: Konsumentenverhalten. 4. Auflage. München: Vahlen.
Kuhlmann, S., Bogumil, J. and Wollmann, H. 2004 (eds.): Leistungsmessung und –vergleich in Politik und Verwaltung. Konzepte und Praxis. Wiesbaden VS Verlag.

Laffin, M. 2008: Local Government Modernisation in England: A Critical Review of the LGMA Evaluation Studies. In: Local Government Studies, 34:1, 109-125.
Landesanstalt für Umweltschutz Baden-Württemberg 1998 (ed.): Umweltmanagement für kommunale Verwaltungen. Leitfaden zur Anwendung der EG-Öko-Audit-Verordnung. Karlsruhe: Landesamt für Umweltschutz Baden-Württemberg.
Landesanstalt für Umweltschutz Baden-Württemberg 2002 (ed.): Öko-Audit in Landesbehörden. Leitfaden zur Einführung und Umsetzung eines Umweltmanagementsystems gemäß EG-Öko-Audit-Verordnung (EMAS). Karlsruhe: Landesamt für Umweltschutz Baden-Württemberg.
Landeshauptstadt Hannover, Amt für Umweltschutz 1998 (ed.): Umwelterklärung `98. Hannover.
Landeshauptstadt Hannover, Amt für Umweltschutz 1999 (ed.): Vereinfachte Umwelterklärung `99. Öko-Audit im Amt für Umweltschutz. Hannover.
Landtag von Baden-Württemberg 2006 (ed.): Drucksache 14/207, 27.07.2006. Stuttgart.
Landtag von Baden-Württemberg 2006a (ed.): Drucksache 14/520, 02.11.2006. Stuttgart.
Lane, J.-E. 2000: New Public Management. London: Routledge.
Lasswell, H. D. 1951: The Policy orientation, in: Lerner, D. and Lasswell, H. D. (eds.): The Policy Sciences: Recent Developments in Scope and Method. Stanford: Stanford University Press, 3-15.
Lasswell, H. D. 1956: The Decision Process. Seven Categories of Functional Analysis. University of Maryland.
Lasswell, H. D. 1968: Language of Politics: Studies in quantitative semantics. Cambridge, Mass.: MIT Press.
Lathrop, D. A. 2003: The Campaign continues. How political consultants and campaign tactics affect public policy. Westport, Conn.: Praeger.
Laux, E. 1999: Erfahrungen und Perspektiven der Kommunalen Gebiets- und Funktionalreformen. In: Wollmann, H. and Roth, R. (eds.): Kommunalpolitik. Opladen: Leske und Budrich, 168-185.

Lepper, M. 1976: Das Ende eines Experiments. Zur Auflösung der Projektgruppe Regierungs- und Verwaltungsreform. In: Die Verwaltung, 4, 478-490.

LfU Baden-Württemberg/LfU Bayern (eds.) 2003: 10 Jahre EMAS, 5 Jahre kommunales Öko-Audit: Bilanz und Perspektiven. Karlsruhe/Augsburg.

Lindblom, C. E. 1990: Inquiry and Change. New Haven: Yale University Press.

Ling, T. 2002: Delivering joined public government in the UK; dimensions, issues and problems. In: Public Administration, 80 (4), 615-642.

Lipsky, M. 1980: Street-Level Bureaucracy. New York: Russell Sage Foundation.

Local Government Association 2006: Local Government Structure. (http://www.lga.gov.uk/Documents/Briefing/LG%20structure%202003.pdf, 05 November 2006).

Löffler, E. 2003: Ökonomisierung ist nicht gleich Ökonomisierung: Die Ökonomisierung des öffentlichen Sektors aus international vergleichender Sicht. In: Harms, P. and Reichard, C.: Die Ökonomisierung des öffentlichen Sektors. Baden-Baden: Nomos, 75-100.

Loew, T. and Clausen, J. 2005: Wie weiter mit EMAS? Schlussfolgerungen vor dem Hintergrund einer Längsschnittanalyse von 1997 bis 2002. Berlin 2005 (http://www.4sustainability.org/downloads/Loew-Clausen2005-Wie-weiter-mit-EMAS.pdf, 22 March 2007).

LUBW Baden-Württemberg (ed.) 2007: Gemeinsamer Arbeitskreis mit den kommunalen Umweltverbänden 2006. (http://www.lubw.baden-wuerttem berg.de/servlet/is/24902/, 12 April 2007), Karlsruhe.

Luhmann, N. 1984: Soziale Systeme. Grundriss einer Allgemeinen Theorie. Frankfurt/Main: Suhrkamp.

Lütkes, S. 1999: Erweiterung der EG-Öko-Audit-Verordnung auf Dienstleistungsunternehmen. In: Schimmelpfenning, L. and Henn, S. (eds.): UmweltManagementn für Handel Banken und Versicheurngen. Verküpfung von Ökologie und Ökonomie durch bewährte Systeme für die Praxis.

Machmer, D. 1995: Leitfaden zur Umsetzung des EG-Öko-Auditsystems in Unternehmen. Offenbach: Umweltinstitut.

Maier, H. 1980: Die ältere Deutsche Staats- und Verwaltungslehre (Polizeywissenschaft). 2. Auflage, München: Beck.

Malek, T. 2000: Die Entstehung der Verordnung. In: Heinelt, H. et. al.: Prozedurale Umweltpolitik der EU. Umweltverträglichkeitsprüfungen und Öko-Audits im Vergleich. Opladen: Leske und Budrich, 67-74.

Malik, F. and Probst, G. 1981: Evolutionäres Management. In: Die Unternehmung. Vol. 35, 121-140.

Martin, S. and Bovaird, A. 2005: Meta-Evaluation of the Local Government Modernisation Agenda: Progress Report on Service Improvement in Local Government (London: ODPM, (http://www.communities.gov.uk/documents/localgovernment/pdf/146058, 08 February 2008).

May, J. P. and Wildavski, A. (eds.) 1978: The Policy Cycle. Beverly Hills: Sage Publishers.

Mayntz, R. 1963: Soziologie der Organisation. Reinbek bei Hamburg: Rowohlt.

Mayntz, R. 1977: Die Implementation politischer Programme: Theoretische Überlegungen zu einem neuen Forschungsgebiet. In: Die Verwaltung, 1, 55. (printed in: Mayntz, R. (ed.): Implementation politischer Programme 1. Empirische Forschungsberichte. Königsstein/Ts: Verlagsgruppe Athenäum).

Mayntz, R. 1980: Gesetzgebung und Bürokratisierung: Wissenschaftliche Auswertung der Anhörung zu den Ursachen einer Bürokratisierung in der öffentlichen Vewaltung. Bonn: Bundesminister des Innern.

Mayntz, R. 1983 (ed.): Implementation politscher Programme 2. Ansätze zur Theoriebildung. Opladen: Westdeutscher Verlag.

Mayntz, R. 1985: Soziologie der öffentlichen Verwaltung. Dritte Auflage. Heidelberg: Müller, Juristischer Verlag.

Mayntz, R. 1987: Politische Steuerung und gesellschaftliche Steuerungsprobleme – Anmerkungen zu einem analytischen Paradigma. In: Ellwein, T. et. al. (eds.): Jahrbuch zur Staats- und Verwaltungswissenschaft 1. Baden-Baden: Nomos, 89-110.

Mayntz, R. 1993: Policy-Netzwerke und die Logik von Verhandlungssystemen. In: Heritier, A. (ed.): Politische Vierteljahresschrift, Sonderheft Policy Analyse, 39-56.

Mayntz, R. and Scharpf, F. W. 1995: Der Ansatz des akteurszentrierten Institutionalismus. In Mayntz, R. and Scharpf, F. W. 1995: Gesellschaftliche Selbstregulierung und politische Steuerung. Frankfurt/Main: Campus 1995, 39-72.

Mayring, P. 2002: Einführung in die qualitative Sozialforschung. Eine Anleitung zu qualitativem Denken. Weinheim: Beltz.

Mayring, P. 2003: Qualitative Inhaltsanalyse. In: Flick, U. et. al. (eds): Qualitative Forschung. Ein Handbuch. Zweite Auflage. Reinbek bei Hamburg: Rowohlt Taschenbuch-Verlag, 468-475.

McCarthy, J. D. and Zald, Mayer N. 1977: Resource Mobilization and Social Movements: A Partial Theory. In: American Journal of Sociology, 82, 6, 1212-1241.

McIntosh, M. and Smith, R. 2000: Die Implementation der Öko-Audit-Verordnung in Großbritannien. In: Heinelt, H. et. al.: Prozedurale Umweltpolitik der EU. Umweltverträglichkeitsprüfungen und Öko-Audits im Vergleich. Opladen: Leske und Budrich, 189-296.

Meyer, C. 1998: Die Effizienz der Kommunalverwaltung. Eine Analyse der Kommunalverwaltungseformdebatte aus sozialökonomischer Perspektive. Baden-Baden: Nomos.

Miller, D. and Friesen, P. H. 1980: Momentum and Revolution in Organizational Adaptation. In: Academy of Management Journal, 23, 591-614.

Mintzberg, H. and McHugh, A. 1985: Strategy Formation in an Adhocracy. In: Administrative Science Quarterly. 30, 160-197.

Möller, H. 2004: Die verfassungsgebende Gewalt des Volkes und die Schranken der Verfassungsrevision: Eine Untersuchung zu Art. 79 Abs. 3 GG und zur verfassungsgebenden Gewalt nach dem Grundgesetz. Berlin: dissertation.de.

Moorstein-Marx, F. 1965 (ed.): Verwaltung. Eine einführende Darstellung. Berlin: Duncker und Humblot.

Naschold, F. 2000: Modernisierung des öffentlichen Sektors im internationalen Vergleich. In: Naschold, F. and Bogumil, J.: Modernisierung des Staates. New Public Management in deutscher und internationaler Perspektive. 2nd ed. Opladen: Leske und Budrich, 27-77.

Nassmacher, H. and Nassmacher, K.-H. 1999: Kommunalpolitik in Deutschland. Opladen: Leske und Budrich.

National Audit Office: Policy Development: Improving Air Quality. London 2001 (http://www.nao.org.uk/publications/nao_reports/01-02/0102232. pdf).

Neumayer, E. and Perkins, R 2004: Europeanisation of the uneven convergence of environmental policy: explaining the geography of EMAS. Environment and Planning C: Government and Policy, 22 (6): 881-897.

Nitschke, P. 1998: Max Weber und die deutsche Verwaltungswissenschaft – eine gescheiterte Annäherung? In: Laux, E. and Teppe, K. (eds.): Der

Neuzeitliche Staat und seine Verwaltung. Beiträge zur Entstehungsgeschichte seit 1700. Stuttgart: Steiner, 163-176.

Nitschke, P. 1999: Die Politik der neuen Unübersichtlichkeit. In: Nitschke, P. (ed.): Die Europäische Union der Regionen. Subpolity und Politiken der Dritten Ebene. Opladen: Leske und Budrich, 9-18.

Nohlen, D. 1998 (ed.): Lexikon der Politik. Band 7: Politische Begriffe. München: Beck.

Norton, Philip 2001: The British Polity. 4th ed. London, New York: Longman.

Öko-Institut (ed.) 1997: Revision der EMAS-Verordnung. Stellungnahme und Änderungsvorschläge zum Vorschlag der Europäischen Kommission vom 10.10.1997, Version vom 04.11.1997. Im Auftrag des Deutschen Naturschutzrings (DNR) für das Europäische Umweltbüro. Darmstadt.

Oevermann, U. et. al. 1979: Die Methodologie einer "objektiven Hermeneutik" und ihre allgemeine forschungslogische Bedeutung in den Sozialwissenschaften. In: Soeffner, H.-G.: (ed.): Interpretative Verfahren in den Sozial- und Textwissenschaften. Stuttgart: Metzler, 352-434.

Oevermann, U. 2002: Klinische Soziologie auf der Basis der Methodologie der objektiven Hermeneutik – Manifest der objektiv hermeneutischen Sozialforschung. Frankfurt am Main, (http://www.ihsk.de/publikationen/Ulrich_Oevermann-Manifest_der_objektiv_hermeneutischen_Sozialforschung.pdf, 17 March 2006).

Offe, C. 1972: Strukturprobleme des kapitalistischen Staates. Aufsätze zur Soziologie. Frankfurt/Main: Suhrkamp.

Office of National Statistics (ed.) 2003: www.statistics.gov.uk/pdfdir/980629-2.htm (29 March 2003).

Orthmann, F. 2002: Effizienzsteigerung im Umweltschutz: das EG-Umweltauditsystem als Instrument anreizorientierter Umweltpolitik. Wiesbaden: Deutscher Universitätsverlag.

Osborne, D. and Gaebler, T. 1994: Reinventing Government. – How the Entrepreneurial Spirit is Transforming the Public Sector. 13th ed. Reading, Mass.: Addison-Wesley.

Osbourne, S. P. and McLaughlin, K. 2008: The study of public management in Great Britain. Public service delivery and its management. In: Kickert, W. (ed.): The Study of Public Management in Europe and the US. London: Routledge, 70-98.

Osner, A. 2001: Kommunale Organisations-, Haushalts- und Politikreform. Ökonomische Effizienz und politische Steuerung. Berlin: E. Schmidt.

Patton, M. Q. 1990: Qualitative evaluation and research methods, 2nd ed. Sevenoaks: Sage.

Pautzke, G. 1989: Die Evolution der organisatorischen Wissensbasis. Bausteine zu einer Theorie des organisatorischen Lernens. München: Kirsch.

Pawson, R. and Tilley, N. 1997: Realistic Evaluation. London: Sage.

Peele, G. 2004: Governing the UK. 4th ed. Oxford: Blackwell.

Peglau, R. 1996: A new Approach to the 5th European Environmental Protection Programme: The Principles, Rules and present Application for the Environmental Management and Audit Scheme (EMAS Regulation). Asia-Pacific Regional Seminar on Environmental Standards, ISO 14001 and the Industry.

Pichel, K. 2002: Innerbetriebliches Ecopreneurship durch Umweltmanagementsysteme? Eine emprische Analyse von Bedingungen umweltbewussten Verhaltens. Berlin: dissertation.de.

Pollitt, C. and Bouckaert, G. 2000: Public Management Reform. A comparative analysis. Oxford: Oxford University Press.

Popper, K. R. and Eccles, J. C. 1989: Das Ich und sein Gehirn. München: Piper.

Prätorius, R. 1997: Theoriefähigkeit durch Theorieverzicht? Zum staatswissenschaftlichen Ertrag der Policy-Studien. In: Benz, A. and Seibel, W. (eds.): Theorieentwicklung in der Politikwissenschaft – eine Zwischenbilanz. Baden-Baden: Nomos, 283-301.

Prittwitz, V. von 1990: Das Katastrophen-Paradox. Elemente einer Theorie der Umweltpolitik. Opladen: Leske und Budrich.

Prittwitz, V. von 1994: Politikanalyse. Opladen: Leske und Budrich.

Probst, G. J. and Büchel, B. S. 1994: Organisationales Lernen. Wettbewerbsvorteil der Zukunft. Wiesbaden: Gabler.

Püttner, G. 2000: Verwaltungslehre. Ein Studienbuch. 3. Auflage. München: Beck.

Pyper, R. 2001: Civil Service management and policy. In: Jones, B. et. al.: Politics UK. 4 th ed., Harlow: Longman, 458-477.

Richards, D. and Smith, M. J. 2002: Governance and Public Policy in the UK. Oxford: Oxford University Press.

Rehbinder, E. and Stewart, R. 1985: Enviromental Protection Policy. Berlin: De Gruyter.

Reichard, C. 1995: Von Max Weber zum „New Public Management" – Verwaltungsmanagement im 20. Jahrhundert. In: Halblützel, P. et. al. (eds.): Umbruch in Politik und Verwaltung. Ansichten und

Erfahrungen zum New Public Management in der Schweiz. Bern: Haupt, 57-79.

Reichard, C. 1996: Die „New Public Management"-Debatte im internationalen Kontext. In: Reichard, C. and Wollmann, H. (eds.): Kommunalverwaltungen im Modernisierungsschub? Basel: Birkhäuser, 241-274

Reinermann, H. 2000: Neues Politik- und Verwaltungsmanagement. Leitbild und theoretische Grundlagen. Speyerer Arbeitshefte 130. Speyer: DHV.

Rhodes, R.A.W. 1994: The hollowing out of the state. In: Political Quarterly, 65, 2, 138-51

Riglar, N. 1995: What is EMAS? In: EG Magazine, June, 5-8.

Riglar, N. 1997: Another year, another EMAS survey. EG Magazine, October, 47-70.

Rindermann, H. 1992: Die Entwicklung der EG-Umweltpolitik von den Anfängen bis 1991. Münster: Lit.

Rist, R. 1990 (ed.): Program Evaluation and the Management of Government. New Brunswick: Transaction Publishers.

Ritz, A. 2003: Evaluation von New Public Management. Bern: Haupt.

Rose, R 1982: Understanding the United Kingdom. London: Longman.

Rose, R. 1991: What is Lesson Drawing? Journal of Public Policy. 11(1): 3-30.

Rothery, B. 1993: BS 7750: Implementing the Environmental Management standards and the EC Eco Management Scheme. Aldershot: Gower.

Sabatier, P. A. 1993: Advocacy-Koalitionen, Policy-Wandel und Policy-Lernen: Eine Alternative zur Phasenheuristik. In: Héritier, A.: (ed.): Policy-Analyse. Kritik und Neuorientierung. Politische Vierteljahresschrift 34, Sonderheft 24, Opladen: Westdeutscher Verlag 1993, 116-148.

Sabatier, P. A. 1999: The Need for Better Theories. In: Sabatier, P. A. (ed.): Theories of the Policy Process. Boulder, Col: Westview Press, 3-17.

Saldern, A. von 1998: Rückblicke. Zur Geschichte der kommunalen Selbstverwaltung in Deutschland. In: Wollmann, H. and Roth, R. (eds.): Kommunalpolitik. Politisches Handeln in den Gemeinden. Bonn: Bundeszentrale für politische Bildung, 23-49.

Schanz, G. 1992: Organisation. In: Frese, E. (ed.): Handwörterbuch der Organisation. 3rd ed. Stuttgart: Poeschel, 1460-1471.

Scharpf, F. W. 1973: Verwaltungswissenschaft als Teil der Politikwissenschaft. In: Scharpf, F. W. (ed.): Planung als politischer Prozess. Aufsätze zur Theorie der planenden Demokratie. Frankfurt/Main: Duncker und Humblot, 9-32.

Scharpf, F. W. 2000: Interaktionsformen. Akteurszentrierter Institutionalismus in der Politikforschung. Opladen: Leske und Budrich.

Schein, E. H. 1985: Organizational Culture and Leadership. San Francisco: Jossey-Bass.

Schmitz, H.-J. 1997: Öko-Audit und Umweltmanagement: Praxiserfahrungen und Aussichten. In: Wasser, Luft und Boden (wlb), 6/1997, 19-24.

Schneider, V. 2008: Komplexität, Politische Steuerung und evidenz-basiertes Policy-Making. In: Janning, F. and Toens, K. (eds.): Die Zukunft der Policy-Forschung: Theorien, Methoden, Anwendungen. Wiesbaden: VS Verlag für Sozialwissenschaften, 55-70.

Schneider, V. and Janning F. 2006: Politikfeldanalyse: Akteure, Diskurse und Netzwerke in der öffentlichen Politik. Wiesbaden: VS Verlag.

Schreyögg, G. and Papenheim, H. 1988: Kooperationsstrategien. Hagen: Fernuniversität Hagen.

Schröter, E. and Wollmann, H. 1997: Public sector reforms in Germany: when and where. A case of ambivalence. In: Hallinon Tutkimus, Administrative Studies, 3, 184-200.

Schütze, F. 1983: Biographieforschung und narratives Interview. In: Neue Praxis, 13 (3), 283-293.

Schwaderlapp, R. 1999: Umweltmanagementsysteme in der Praxis: Qualitative empirische Untersuchung über die organisatorischen Implikationen des Öko-Audits. München: Oldenbourg.

Schwalb, L. and Walk H. (eds.) 2007: Local Governance – mehr Transparenz und Bürgernähe? Wiesbaden: VS, Verlag für Sozialwissenschaften.

Schwanke, K. and Elbinger, F. 2006: Politisierung und Rollenverständnis der deutschen Administrativen Elite 1970 bis 2005 – Wandel trotz Kontinuität. In: Bogumil, J., Jann, W. and Nullmeier, F. (eds.): Politik und Verwaltung. PVS Sonderheft 37/2006, 228-249.

Shrivastava, P. A. 1983: Typology of Organizational Learning Systems. In: Journal of Management Studies, 20 (1), 7-38.

Siedentopf, H. 1976: Einleitung. In: Siedentopf, H. (ed.): Verwaltungswissenschaft. Darmstadt: Wissenschaftliche Buchgesellschaft, 1-17.

Slapper, G. and Kelly, D. 2001: The English Legal System. Fifth ed. London: Cavendish.

Smirchich, L. 1983: Concepts of culture and organizational analysis. In: Administrative Science Quarterly, 28, 330-358.

Spitzer, M. 1998: Bürgeraktivierung und Verwaltungsmodernisierung. In: Bademer, S. von et. al. (eds.): Handbuch zur Verwaltungsreform. Opladen: Leske und Budrich, 131-139.

Staggenborg, S. 2008: Social Movements. Oxford: Oxford University Press.

Stevens, A. 2003: Politico's Guide to Local Government. London: Politico.

Stockmann, R. (ed.) 2000: Evaluationsforschung. Opladen: Leske und Budrich.

Stoker, G. and Mossberger, K. 1995: The Post Fordist Licals State: The Dynamics of its Development, In: Steward, J. and Stoker, G.: Local Government in the 1990s. London: Macmillan, 210-227.

Stoker, G. 2004: Transforming Local government: From Thatcher to New Labour. Houndsmill: Palgrave Macmillan.

Sullivan, H. and Gillanders, G. 2005: Stretched to the limit? The impact of local public service agreements on service improvements and central-local relation. In: Local Government Studies, 31 (5), 555-574.

Tarrow, S. 1998: Power in Movement: Social Movement and Contentious Politics. 2nd ed. Cambride: Cambridge University Press.

Teichert, V. 2000: Umweltmanagementsysteme in Schulen. Heidelberg: FEST.

The Stationery Office: Local Government Act 2000. London. (http://www.hmso.gov.uk/acts/acts2000/20000022.htm, 16 June 2003).

Thieme, W. 1984: Verwaltungslehre. 4. Auflage. Köln: Heymanns.

Thom, N. and Ritz, A. 2004: Public Management. Innovative Konzepte zur Führung im Öffentlichen Sektor. 2. Auflage. Wiesbaden: Gabler.

Töller, A. E. and Heinelt, H. 2003: The Negotiation and Renegotiation of a European policy tool. In: Heinelt, H. and Smith, R. (eds.): Sustainability, Innovation and Participartory Governance. A Cross-National Study of the EU Eco-Management and Audit Scheme. Aldershot: Ashgate, 23-52.

Toens, K. and Landwehr, C. 2008: Imitation, Bayesianisches Updating und Deliberation: Strategien und Prozesse des Politiklernens im Vergleich. Janning, F. and Toens, K. (eds.): Die Zukunft der Policy-Forschung: Theorien, Methoden, Anwendungen. Wiesbaden: VS Verlag für Sozialwissenschaften, 71-87.

Türk, K., Lemke, T. and Bruch, M. 2002: Organisation in der modernen Gesellschaft. Eine historische Einführung. Wiesbaden: Westdeutscher Verlag.

Umweltbundesamt 1999 (ed.): EG-Umweltaudit in Deutschland. Erfahrungsbericht 1995 bis 1998. Berlin: Umweltbundesamt.

Umweltbundesamt (ed.) 1998: Umweltmanagement in der Praxis. Berlin: Umweltbundesamt.

Umweltbundesamt 2007 (ed.): Umweltmanagementsysteme weltweit/ Environmental management systems world wide (sic) as January 2006 (sic) (Peglau-Liste) (http://www.umweltbundesamt.de/uba-info-daten/daten/ums-welt.htm, 20 March 2007).

Umweltgutachterausschuss 2005 (ed.): Hintergrundpapier des Umweltgutachterausschusses zur Frage der 63. UMK zu einem „zukunftsfähigen EMAS" in Deutschland. (Beschluss der 37. Plenumssitzung am 29. September 2005 in Berlin) Berlin (http://www.uga.de/downloads/ZukunftEMAS_Hintergrundpapier.pdf, 22 March.2007).

Umweltgutachterausschuss 2006 (ed.): Verwaltungserleichterungen für EMAS-Teilnehmer in den Umweltpakten, Erlassen und sonstigen Regelungen der Bundesländer. Berlin 2006. (http://www.emas.de/datenbank/Verwaltungserleichterungen_Stand_060310.pdf, 20 March 2007).

Umweltgutachterausschuss 2007 (ed.): System of Accreditation, Supervision and Registration according to EMAS II. (http://www.uga.de/down loads/Orga-EMAS-eng.pdf, 7 April 2007).

Umweltgutachterausschuss 2008 (ed.): Der neue EMAS III-Entwurf der Europäischen Kommission. In: EMASaktuell, 17, Berlin: Umweltgutachterausschuss, (http://www.emas.de/datenbank/EMAS aktuell-nr17-sept2008.pdf, 1 October 2008).

Umwelt kommunale ökologische Briefe (ed.): Im Konvoi erfolgreich. In: Umwelt kommunale ökologische Briefe, 17/2001, 8.

Wagener, F. 1979: Der Öffentliche Dienst im Staat der Gegenwart. In: Veröffentlichungen der Vereinigung der deutschen Staatsrechtslehrer 38, Berlin: De Gruyter, 215-266.

Wahren, H. K. 1996: Das lernende Unternehmen. Theorie und Praxis des organisationalen Lernens. Berlin: de Gruyter.

Walter-Busch, E. 1996: Organisationstheorien von Weber bis Weick. Berlin: Fakultas.

Waskow, S. 1994: Betriebliches Umweltmanagement. Anforderungen nach der Audit-Verordnung der EG. Heidelberg: Müller.

Weale, A. 1999: European Policy by Stealth. The Dysfunctionality of Functionalism. In: Environment and Planning C: Government and Policy 17 (1), 37-52.

Weale, A. et al. 2000: Environmental governance in Europe: an ever closer union? Oxford: Oxford University Press.

Weber, 1980: Wirtschaft und Gesellschaft. Grundriss der verstehenden Soziologie. 5. Auflage. Tübingen: Mohr.

Wegener, A. 2002: Die Gestaltung kommunalen Wettbewerbs. Strategien in den USA, Großbritannien und Neuseeland. Berlin: Edition Sigma.

Wegrich, K. et. al. 1997: Kommunale Verwaltungspolitik in Ostdeutschland. Basel: Birkhäuser.

Wendisch, N. 2004: Das Leitbild und seine Rolle für das Lernen in Organisationen: die Möglichkeit des EMAS für eine leitbildzentrierte Organisationsentwicklung. 2nd ed. München: ökom-Verlag. Wepler, C. 1999: Europäische Umweltpolitik. Die Umweltunion als Chance für die materielle und institutionelle Weiterentwicklung der europäischen Integration. Marburg: Metropolis Verlag.

Williamson, O. E. 1987: The economic institutions of capitalism: firms, markets relational contracting. New York: Free Press.

Wilson, D. and Game, C. 2002: Local Government in the United Kingdom. Third ed., Basingstoke: Palgrave.

Windhoff-Héritier, A. 1987: Policy-Analyse. Eine Einführung. Frankfurt/ Main: Campus Verlag.

Windhoff-Héritier, A. (ed.) 1993: Policy-Analyse: Kritik und Neuorientierung. Politische Vierteljahresschrift, Sonderheft 24. Opladen: Westdeutscher Verlag.

Wiswede, G. 1991: Soziologie. Ein Lehrbuch für den wirtschafts- und sozialwissenschaftlichen Bereich. Landsberg am Lech: Verlag moderne Industrie.

Wölk, J. 2002: Die Umsetzung von Richtlinien der Europäischen Gemeinschaft. Eine rechtsvergleichende Untersuchung zum Recht der Bundesrepublik Deutschland, der französischen Republik und des Vereinigten Königreichs. Baden-Baden: Nomos.

Wolff, R. 1982: Der Prozess des Organisierens. Zu einer Idee des organisationalen Lernens. Spardorf: Wilfer.

Wollmann, H. 1983: Implementation durch Gegenimplementation „von unten“. Das Beispiel der Wohnungspolitik. In: Mayntz, R. (ed.): Implementation politischer Programme II. Ansätze zur Theoriebildung. Opladen: Westdeutscher Verlag, 168-196.

Wollmann, H. 1991: Implementationforschung/Evaluationsforschung. In: Nohlen, D. (ed.): Wörterbuch Staat und Politik. München: Beck, 235-238.

Wollmann, H. 1996: Verwaltungsmodernisierung: Ausgangsbedingungen, Reformanläufe und aktuelle Modernisierungsdiskurse. In: Reichard, C. and Wollmann, H. (eds.): Kommunalverwaltung im Modernisierungsschub? Basel: Birkhäuser, 1-49.

Wollmann, H. 2000: Comparing Institutional Development in Britain and Germany: (Persistent) Divergence or (Progressing) Convergence? In: Wollmann, H. and Schröter, E. (eds.): Comparing Public Sector Reform in Britain and Germany. Aldershot: Ashgate, 1-26.

Wollmann, H. 2000: Local Government Modernisation in Germany: Between Incrementalism and Reform Waves. In: Public Administration, 78 (4), 915-936.

Wollmann, H. 2003: Kontrolle in Politik und Verwaltung: Evaluation, Controlling und Wissensnutzung. In: Schubert, K. and Bandelow, N. C.: Lehrbuch der Politikfeldanalyse. München: Oldenbourg, 335-360.

Würth, S. 1993: Umwelt-Auditing: Die Revision im ökologischen Bereich als wirksames Überwachungsinstrument für die ökologische Unernehmung. St. Gallen: HSG.

Zapf, W. 1969: Theorien des sozialen Wandels. Köln: Kiepenheuer und Witsch.

Appendices

Chapter 5: Appendix 1a and 1b: Interview guidelines in English and German

Appendix Ia: Interview Guideline (in English)

Feb. 05, 2005

Main Question
How was EMAS introduced and implemented and how would you judge the introduction, the implementation and the course of this process?

In general:
- My idea is to ask questions about the three layers of organisation: individual, group and organisation
- Interview is done on the grounds of the concept of the policy cycle as well as with theories of organisational learning in mind. Are there changes in behaviour, changes in goals or changes of strategies?
- Guideline-orientated expert interview

1. General Questions

Title, job description within organisation, function regarding EMAS

2. Problem Perception

How did you get to know about EMAS? Who informed you about EMAS?
(How does your administration perceive environmental problems?)

3. Agenda Setting

How was EMAS brought on the agenda? How did your organisation begin with EMAS?

4. Policy Formulation

Who participated at the formulation of the environmental policy?
How did you develop the goals as described in the environmental policy and/or in the environmental declaration? How long did this process of formulation of the environmental policy take?
How would you describe this process?

5. Decision Making

Who decided to implement EMAS?
What were the reasons for this decision?

6. Policy Implementation

How was EMAS implemented within your organisation? Who was actively involved in the implementation process? What were the intra-organisational as well as the external reactions like towards the implementation of EMAS?
Which learning processes or changes of behaviour did you find with your colleagues?

7. Evaluation

How would you assess the work with the environmental programme? What was successful, what a failure or which effects did occour?
How would you assess EMAS in total?

In General – About learning processes:
Do you see changes in behaviour of staff members caused by EMAS (single loop)?
Do you see changes of goals of staff members caused by EMAS (double loop)?
Do you see changes in the organisational strategy caused by EMAS (deuteron learning?)

8. Policy Termination or New formulation of Policy

Revalidation:
Did your organisation decided for a revalidation? What were the reasons for this decision?
Did your organisation decide to introduce a different management system? What were the reasons for this?

If applicable:
When did you terminate the EMAS-process?

Appendix Ib: Interview Guideline (in German)

Fragestellung:
Wie wurde EMAS eingeführt sowie umgesetzt und wie beurteilen Sie die Einführung, die Durchführung und den bisherigen Verlauf von EMAS?

Grundsätzlich:
- Idee, nach drei Ebenen zu fragen: Individuum, Gruppe und Organisation
- Befragung vor dem Hintergrund des Konzepts des Policy-Cycle sowie von Theorien zum Lernen von Organisationen: Veränderung von Verhalten, Veränderung von Zielen, Veränderung von Strategien?
- Leitfadenorientiertes Experteninterview

1. Allgemeines

Titel, Aufgabe innerhalb der Organisation, Funktion bei EMAS

2. Problemwahrnehmung (Problem Perception)

Wie sind Sie auf EMAS aufmerksam geworden? Wer hat Sie auf EMAS aufmerksam gemacht?
(Wie werden Umweltprobleme in Ihrer Verwaltung wahrgenommen?)

3. Thematisierung (Agenda Setting)

Wie kam EMAS auf die Tagesordnung? Wie wurde mit EMAS begonnen?

4. Politikformulierung (Policy Formulation)

Wer war an der Formulierung der Umweltpolitik maßgeblich beteiligt?
Wie sind Sie auf die in der Umweltpolitik / Umwelterklärung genannten Ziele gekommen?
Wie lange hat der Prozess der Formulierung der Umweltpolitik gedauert?
Wie ist der Prozess abgelaufen?

5. Entscheidung (Decision Making)

Wer hat über den Einsatz von EMAS entschieden?
Gründe für die Entscheidung / Gründe für die Einführung von EMAS?

6. Politikvollzug (Implementation)

Wie gestaltete sich die Umsetzung der EMAS-Richtlinie in Ihrer Organisation? Wer war maßgeblich beteiligt?
Wie reagierte Ihr Umfeld, d.h. sowohl verwaltungsintern als auch organisationsextern auf die Umsetzung von EMAS?
Welche Lernprozesse / Verhaltensänderungen gab es bei den Mitarbeitern?

7. Ergebnisbewertung (Evaluation)

Wie bewerten Sie die Bearbeitung des Umweltprogramms? Welche Erfolge / Misserfolge / Effekte gab es durch EMAS?
Wie bewerten Sie den Einsatz von EMAS insgesamt?

Generell – Lernprozesse betreffend:
Hat sich das Verhalten der Mitarbeiter verändert? (single loop)
Haben sich die Ziele verändert? (double loop)
Haben sich Strategien verändert? (deuteron learning)

8. Politikneuformulierung oder –beendigung (Termination)

Revalidierung
Hat sich Ihre Organisation für eine Revalidierung entschieden? Gründe?
Hat sich Ihre Organisation für die Einführung eines anderen Managementsystems entschieden? Gründe?

Sofern zutreffend:
- **Warum wurde EMAS aufgegeben?**

Aktuelle Probleme moderner Gesellschaften
Contemporary Problems of Modern Societies

Herausgegeben von / Edited by Peter Nitschke und Corinna Onnen-Isemann

Bd./Vol. 1 Corinna Onnen-Isemann / Vera Bollmann: Studienbuch Gender & Diversity. Eine Einführung in Fragestellungen, Theorien und Methoden. 2010.

Bd./Vol. 2 Stephan Sandkötter (Hrsg.): Bildungsarmut in Deutschland und Brasilien. 2010.

Bd./Vol. 3 Astrid Freudenstein: Die Machtphysikerin gegen den Medienkanzler. Der Gender-Aspekt in der Wahlkampfberichterstattung über Angela Merkel und Gerhard Schrö der. 2010.

Bd./Vol. 4 Matthias Bublitz: Gegliederter Universalismus. Politische Philosophie und ihre Ten denzen in der bundesdeutschen Parteienprogrammatik. 2010.

Bd./Vol. 5 Martin Jungwirth: Environmental Management Systems in Local Public Authorities. A Comparative Study of the Introduction and Implementation of EMAS in the United Kingdom and Germany. 2011.

www.peterlang.de